Name _____ Class Meeting Day ____ Class Time _____

Skull – External Details

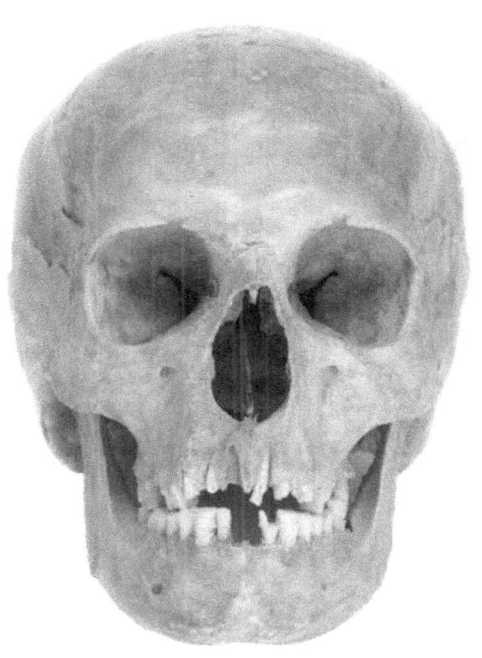

Skull – Lateral Aspect

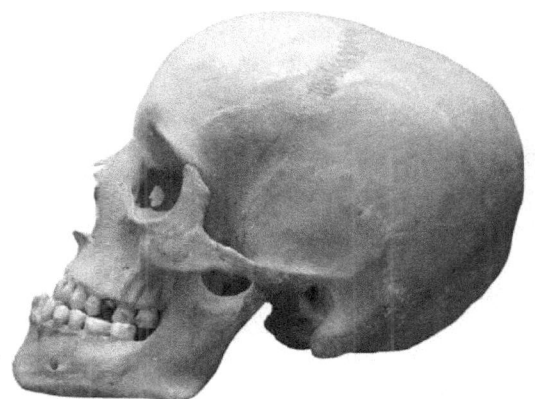

Name _____ Class Meeting Day _____ Class Time _____

Skull – Ventral Aspect (mandible removed)

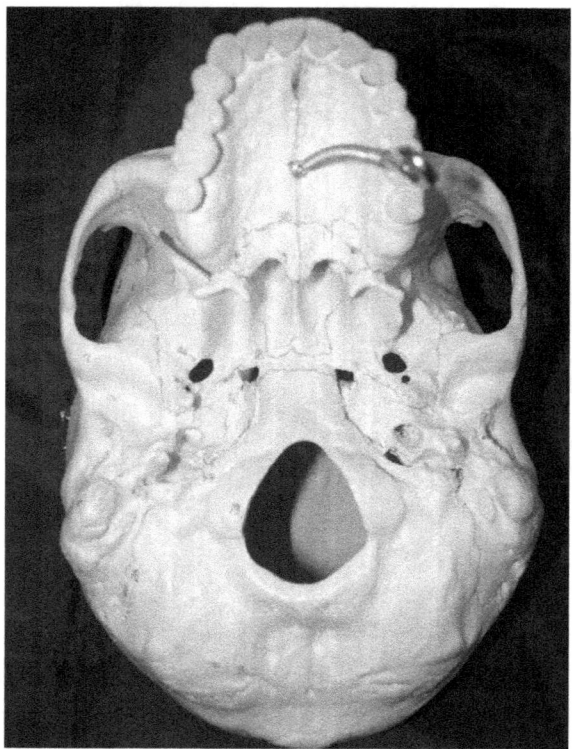

Skull Interior

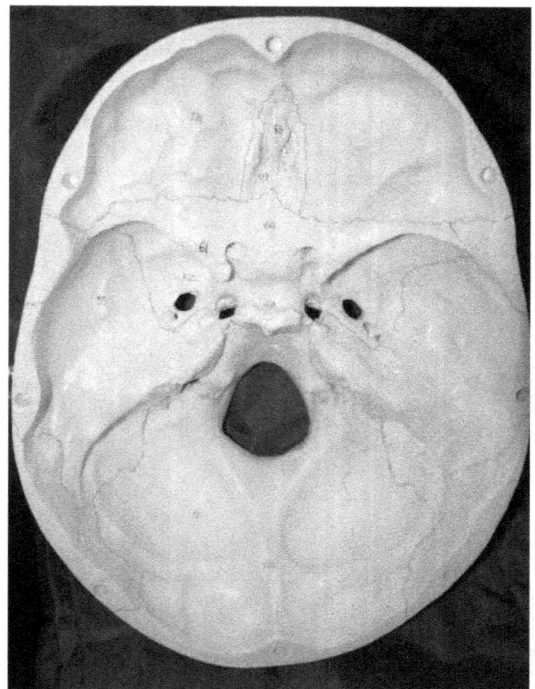

Name _____ Class Meeting Day ____ Class Time _____

Vertebral Column   Cervical Vertebrae (close-up)

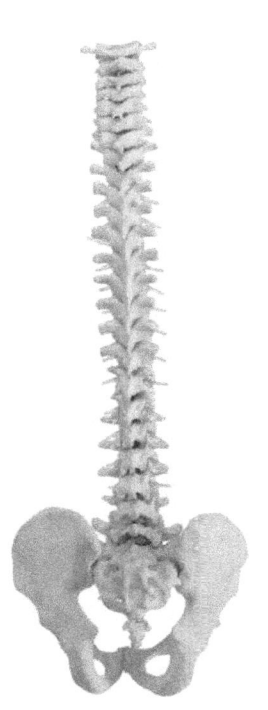
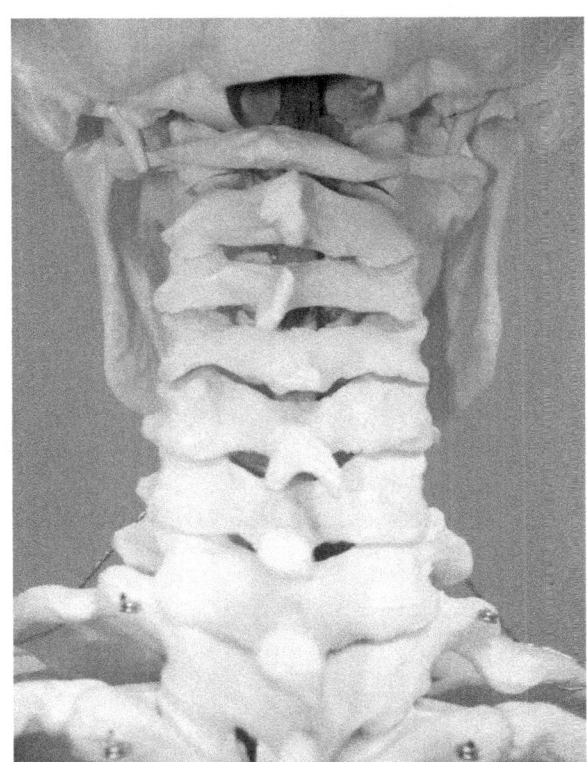

Thoracic Vertebrae   Lumbar Vertebrae

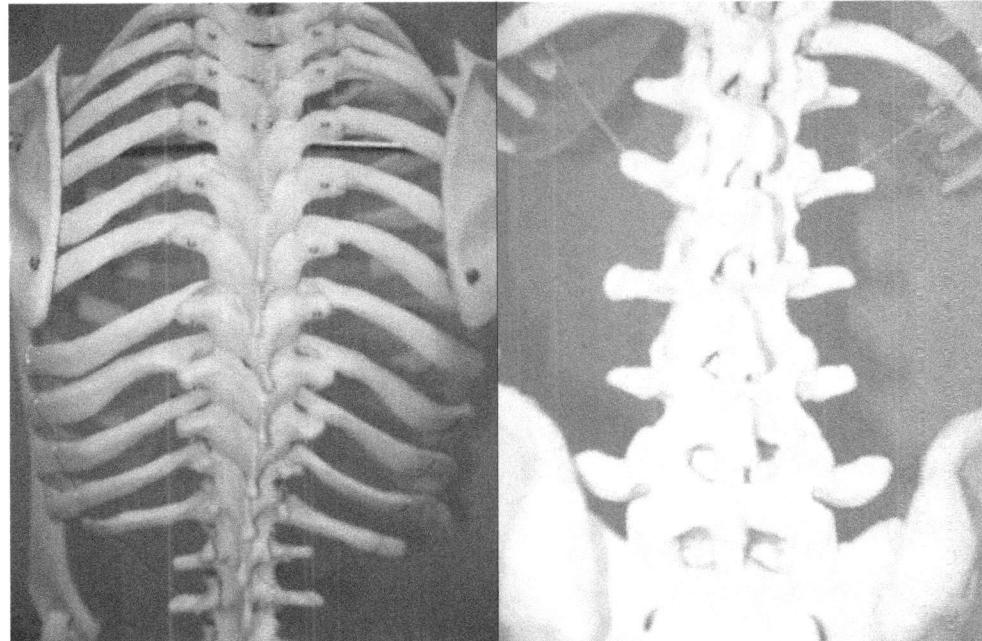

Name _____ Class Meeting Day _____ Class Time _____

Pelvic Girdle (anterior aspect)

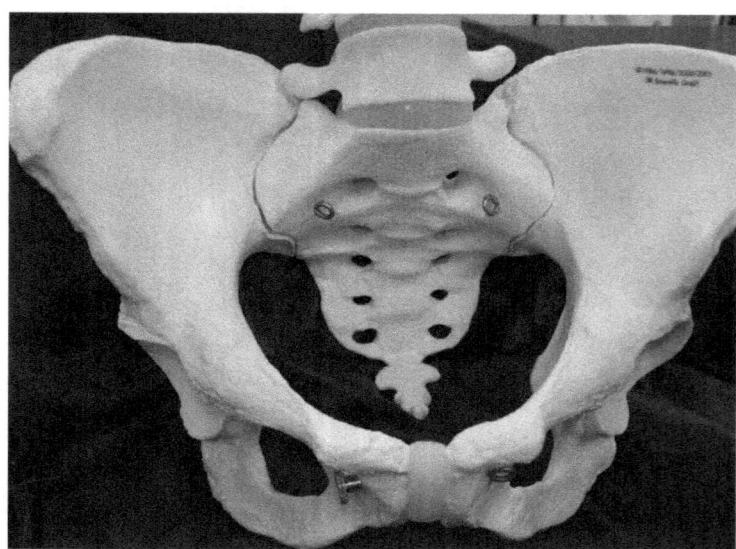

Pelvic Girdle (posterior aspect)

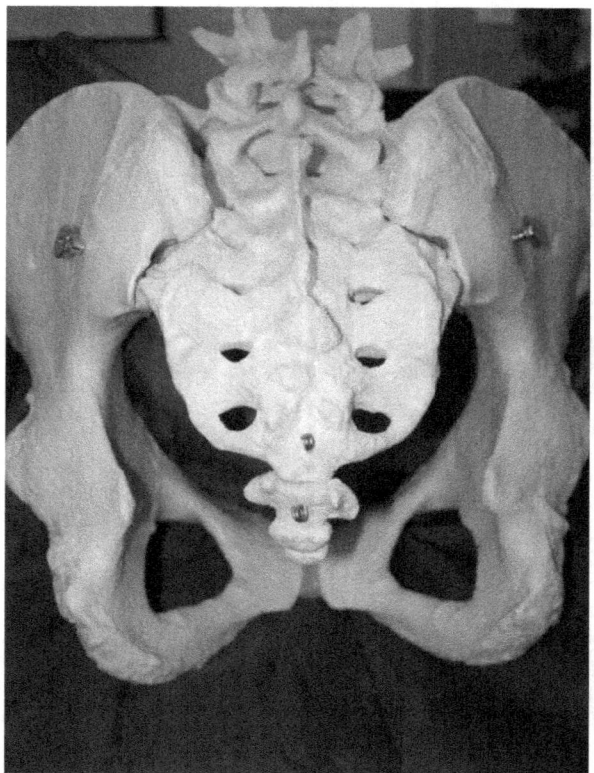

Name _____ Class Meeting Day _____ Class Time _____

Pelvic Girdle (lateral aspect)

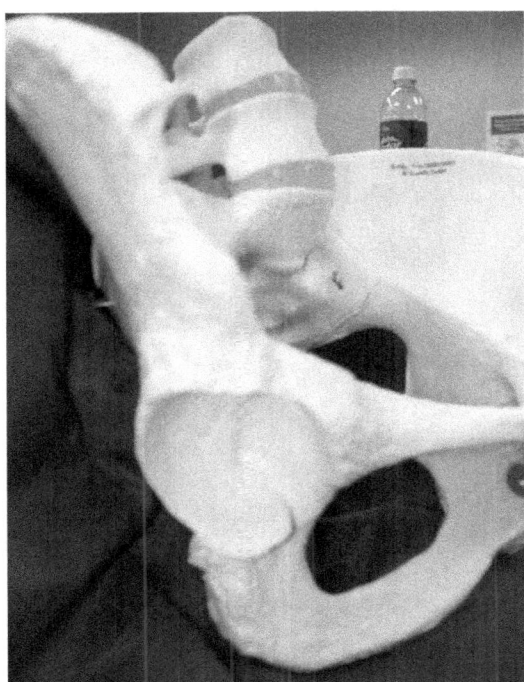

V. Analysis

A. Follow-up Questions

1. What is the largest foramen in the skull?

2. How can you easily distinguish between the radius and the ulna?

3. How can you easily distinguish between the tibia and the fibula?

4. What does articulate mean? (as in the two bones "articulate")

5. How many of each type of vertebrae are in the spine – Cervical? Thoracic? Lumbar?

Name _____ Class Meeting Day _____ Class Time _____

6. What are some differences that you observe in the spines of cervical, thoracic, and lumbar vertebrae?

7. What is the main difference between true and false ribs?

B. Conclusion

1. State in your own words why bone markings are significant.

2. State what you have learned as a result of this laboratory exercise.

Skeletal Key

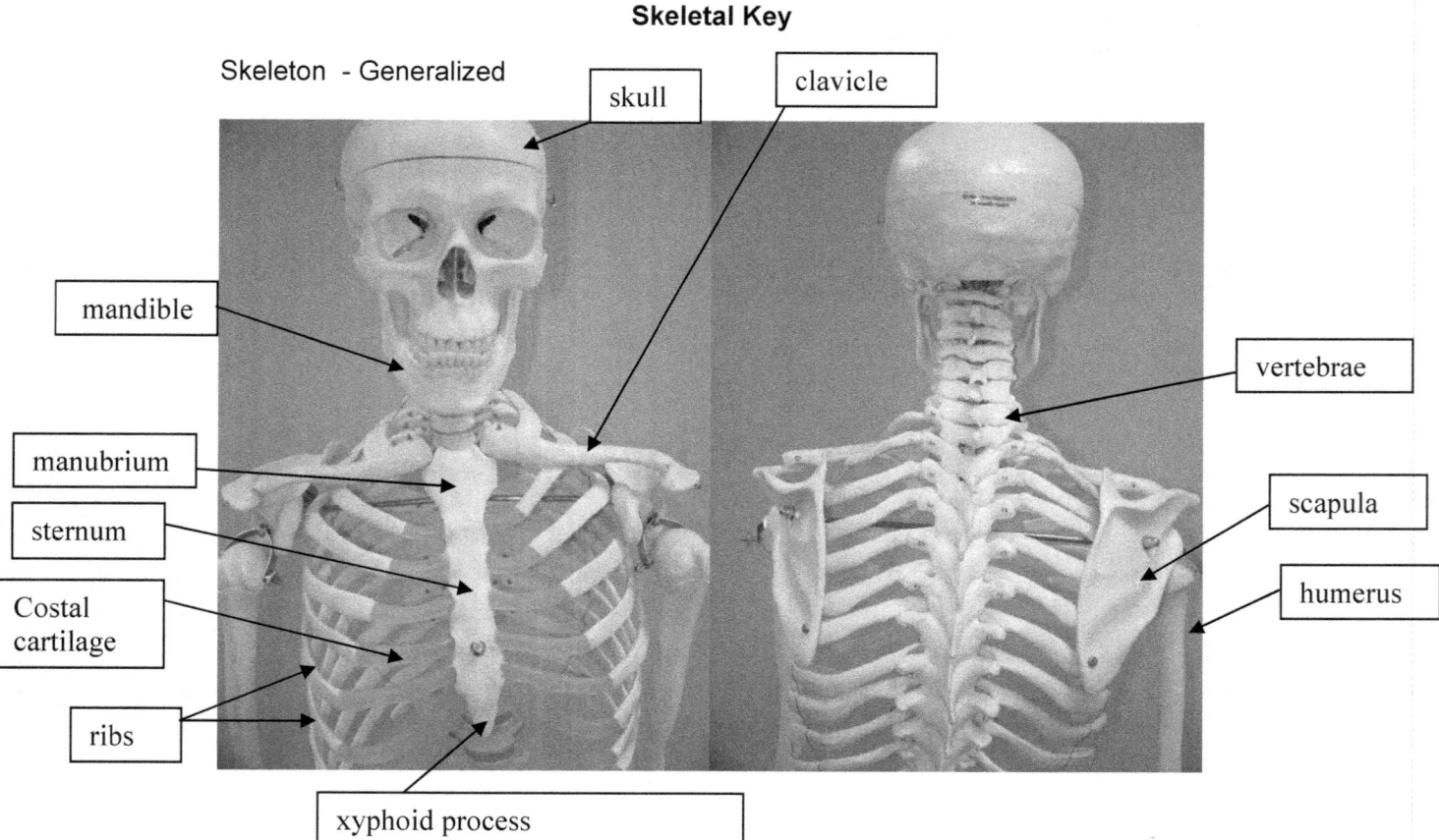

Name _____ Class Meeting Day _____ Class Time _____

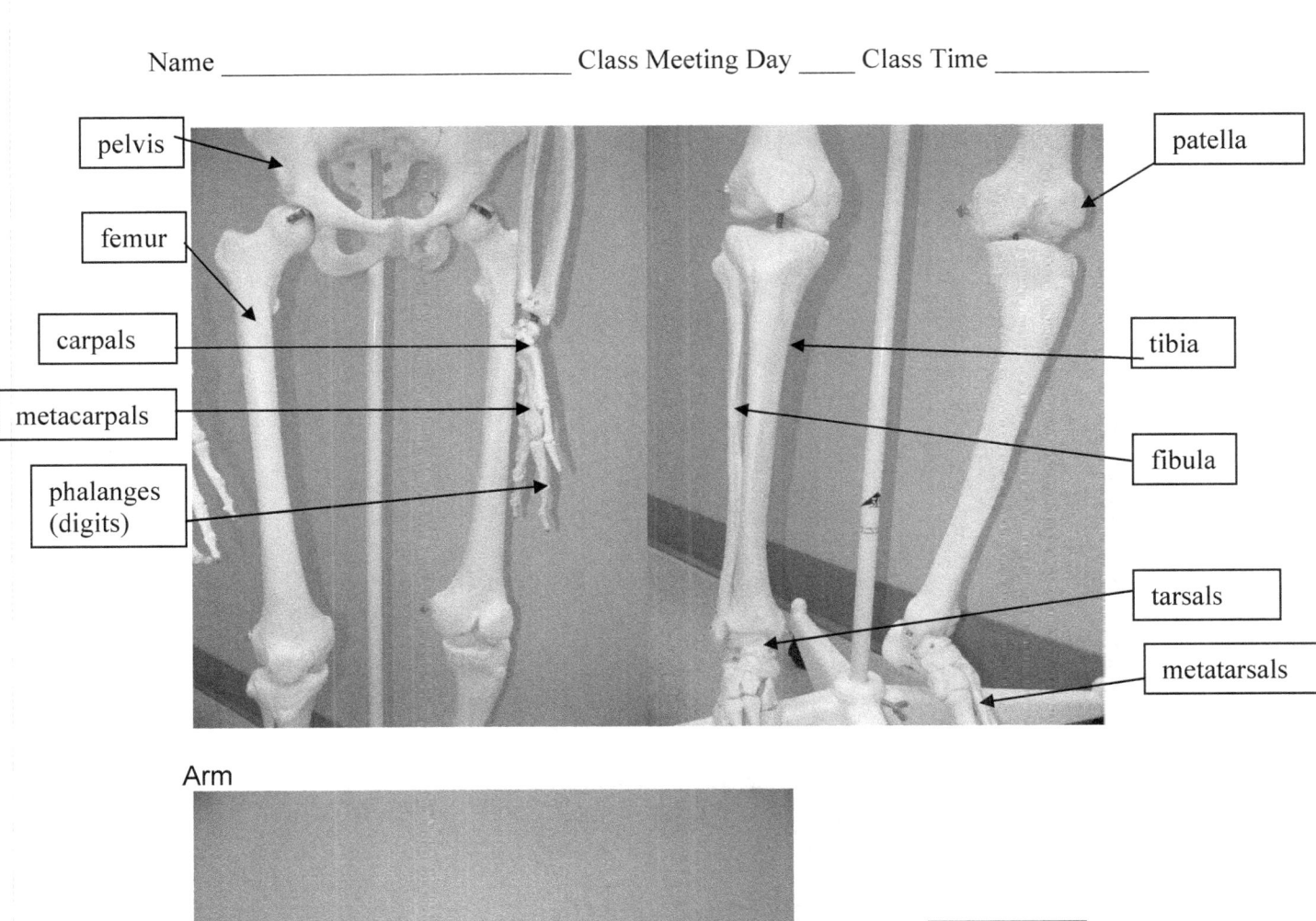

Arm

Name _____ Class Meeting Day ____ Class Time _____

Pectoral Girdle (Posterior Aspect)

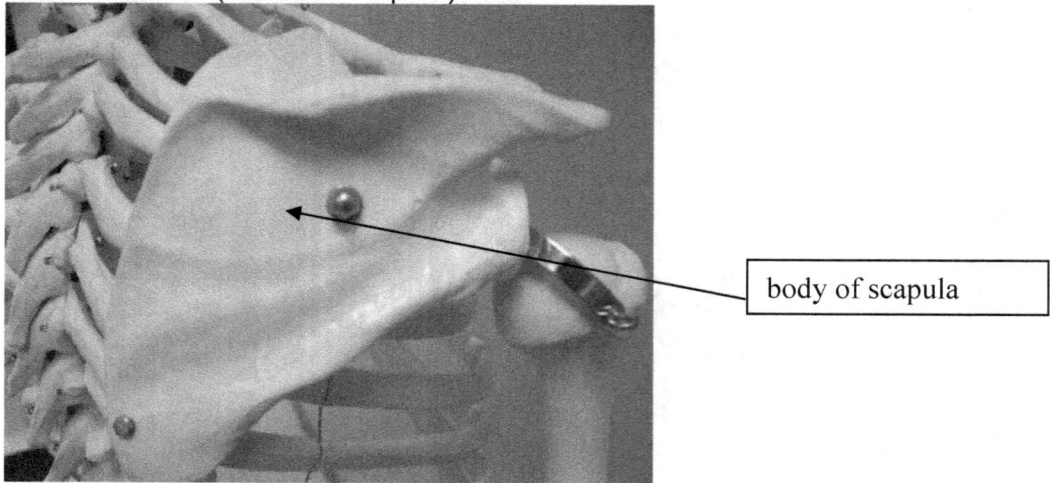

body of scapula

Pectoral Girdle - Anterior Aspect

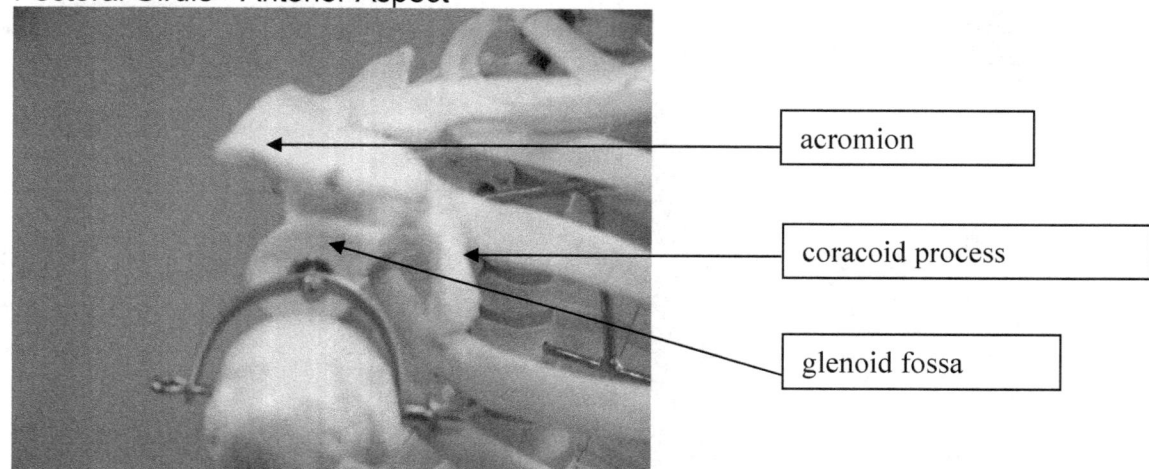

acromion

coracoid process

glenoid fossa

Name _____ Class Meeting Day _____ Class Time _____

Skull – External Details

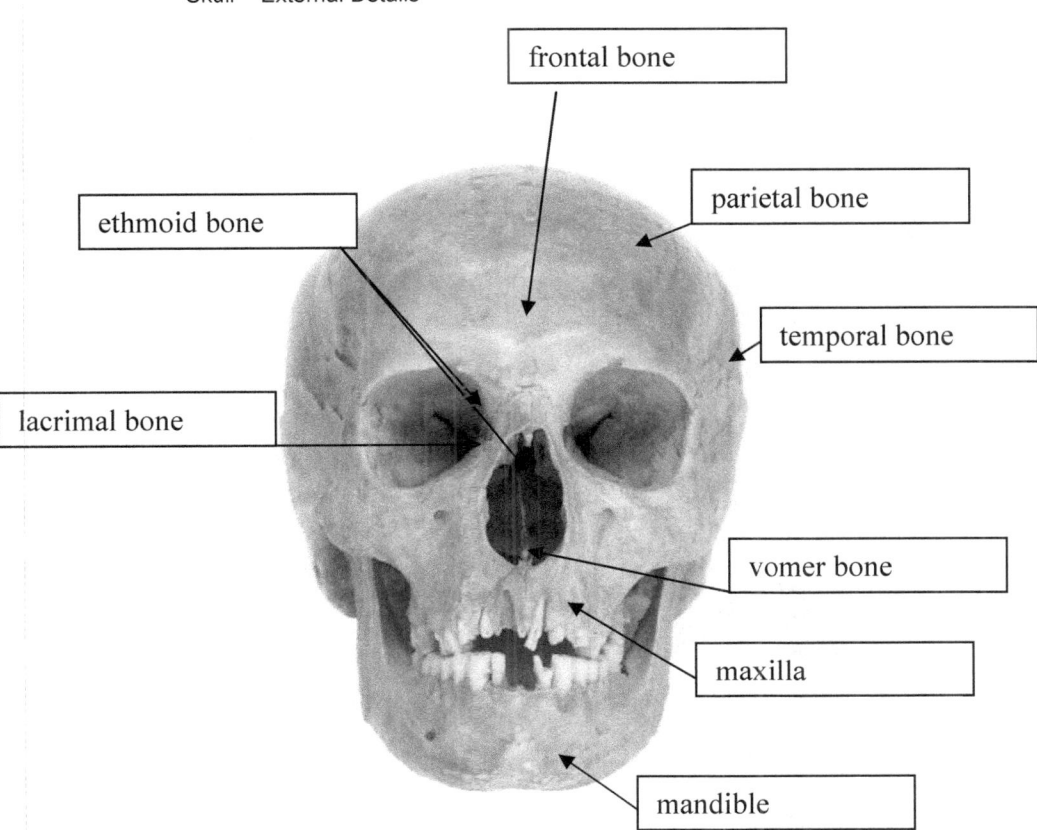

Name _____ Class Meeting Day _____ Class Time _____

Skull – Lateral Aspect

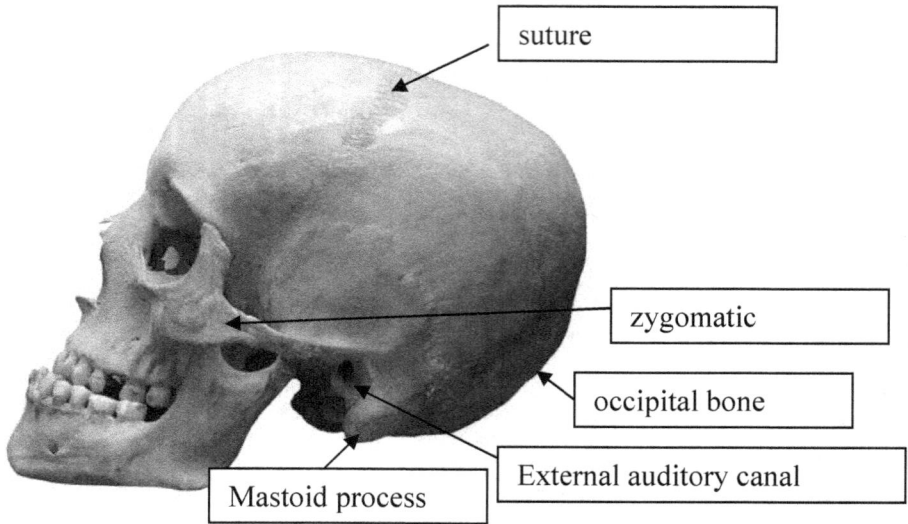

Skull – Ventral Aspect (mandible removed)

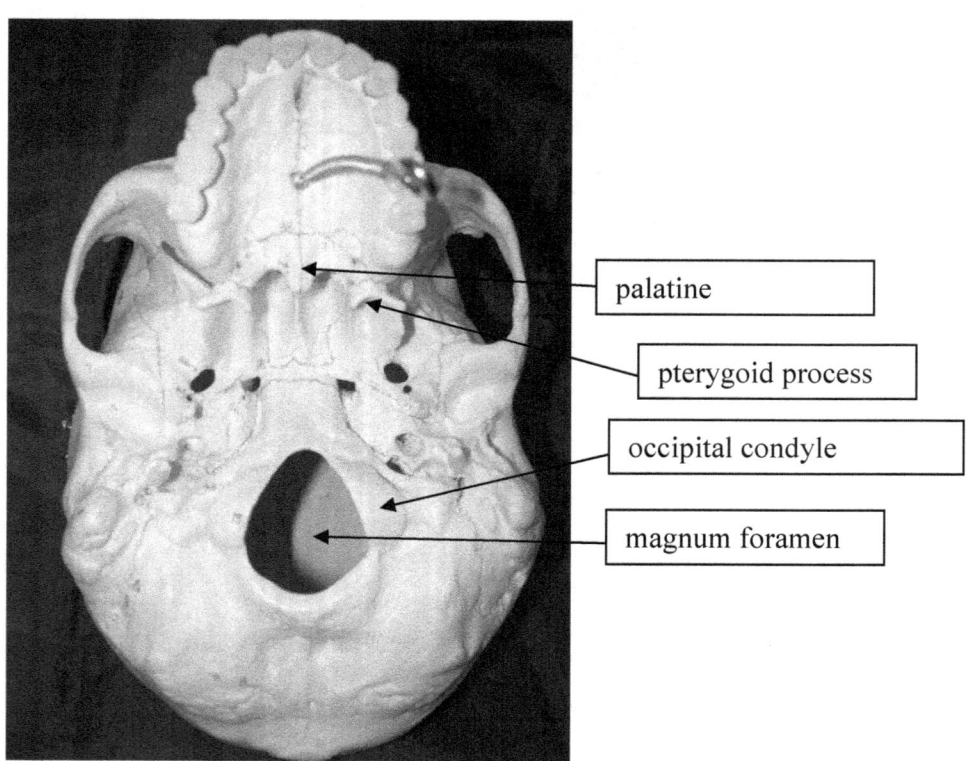

Name _____ Class Meeting Day ____ Class Time _____

Skull Interior

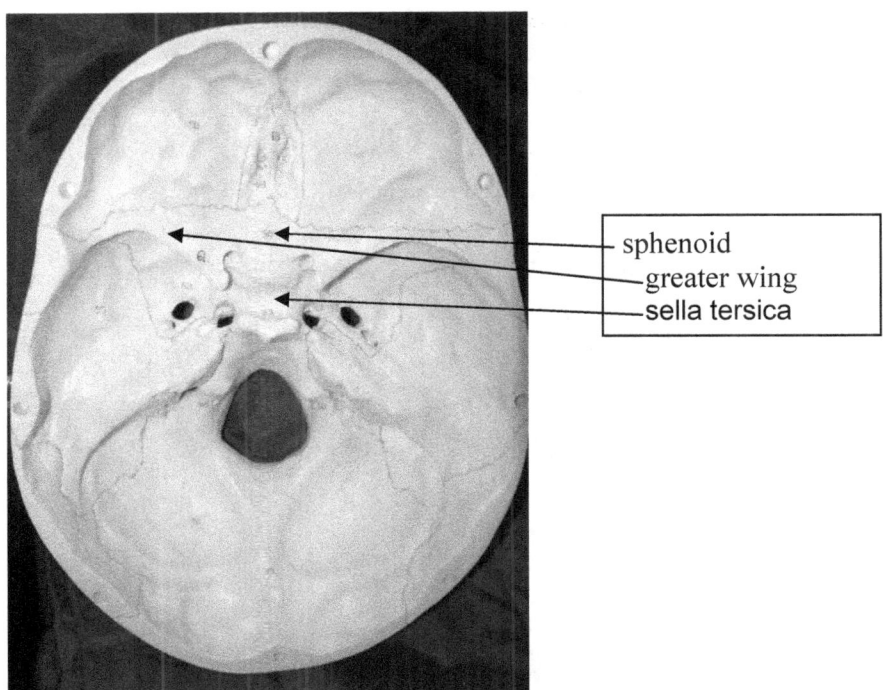

- sphenoid
- greater wing
- sella tersica

Name _____ Class Meeting Day _____ Class Time _____

Vertebral Column

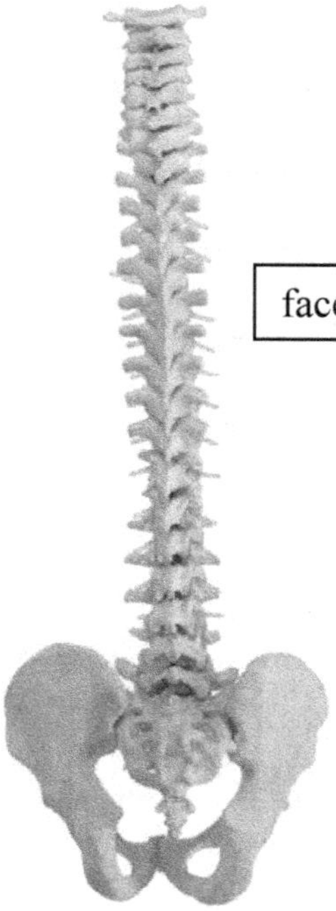

Cervical Vertebrae (close-up)

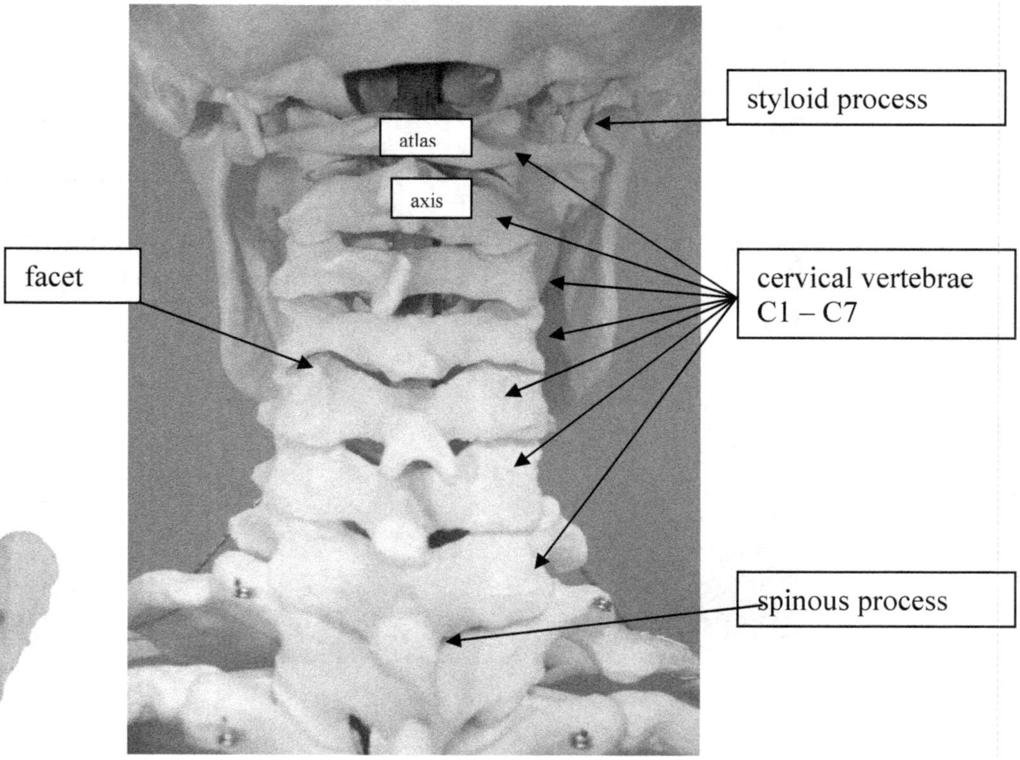

- styloid process
- atlas
- axis
- facet
- cervical vertebrae C1 – C7
- spinous process

Thoracic Vertebrae

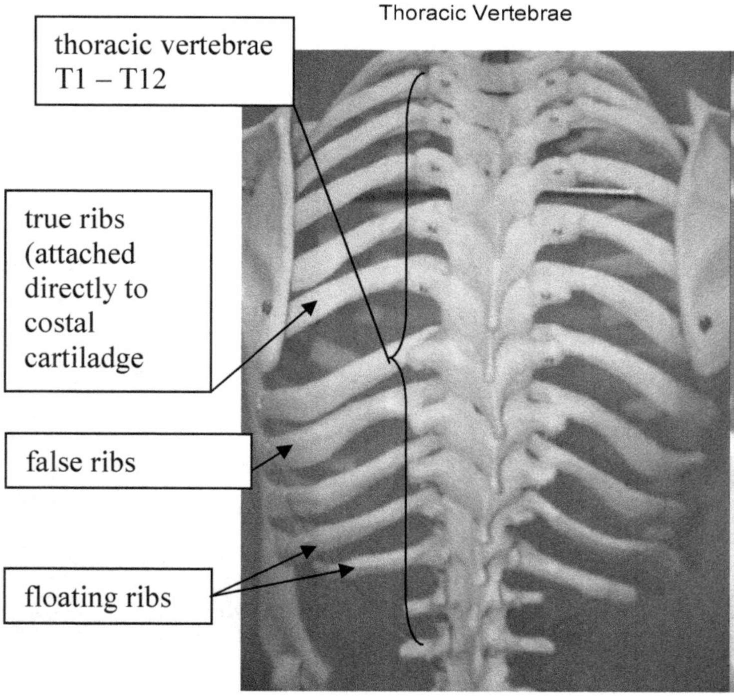

- thoracic vertebrae T1 – T12
- true ribs (attached directly to costal cartiladge)
- false ribs
- floating ribs

Lumbar Vertebrae

- lumbar vertebrae L1 – L5

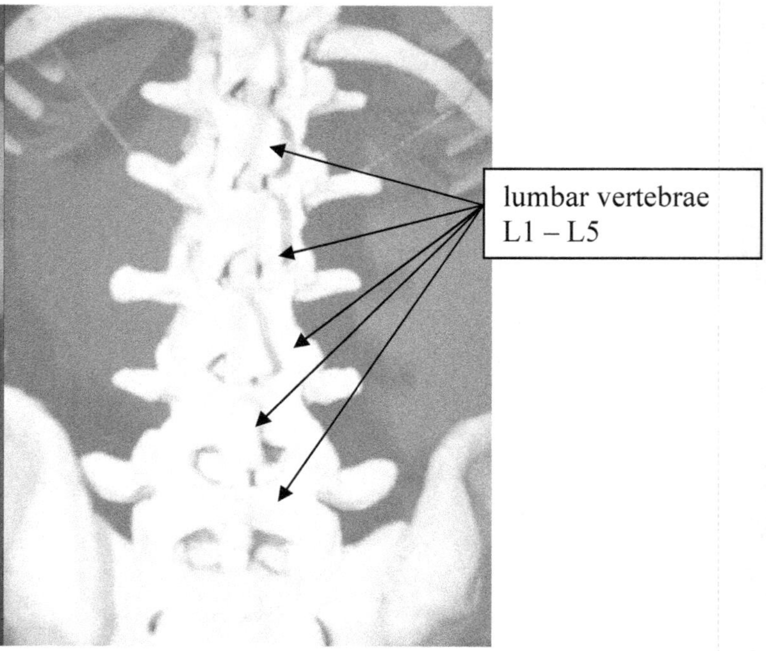

Name _____ Class Meeting Day ____ Class Time _____

Pelvic Girdle (anterior aspect)

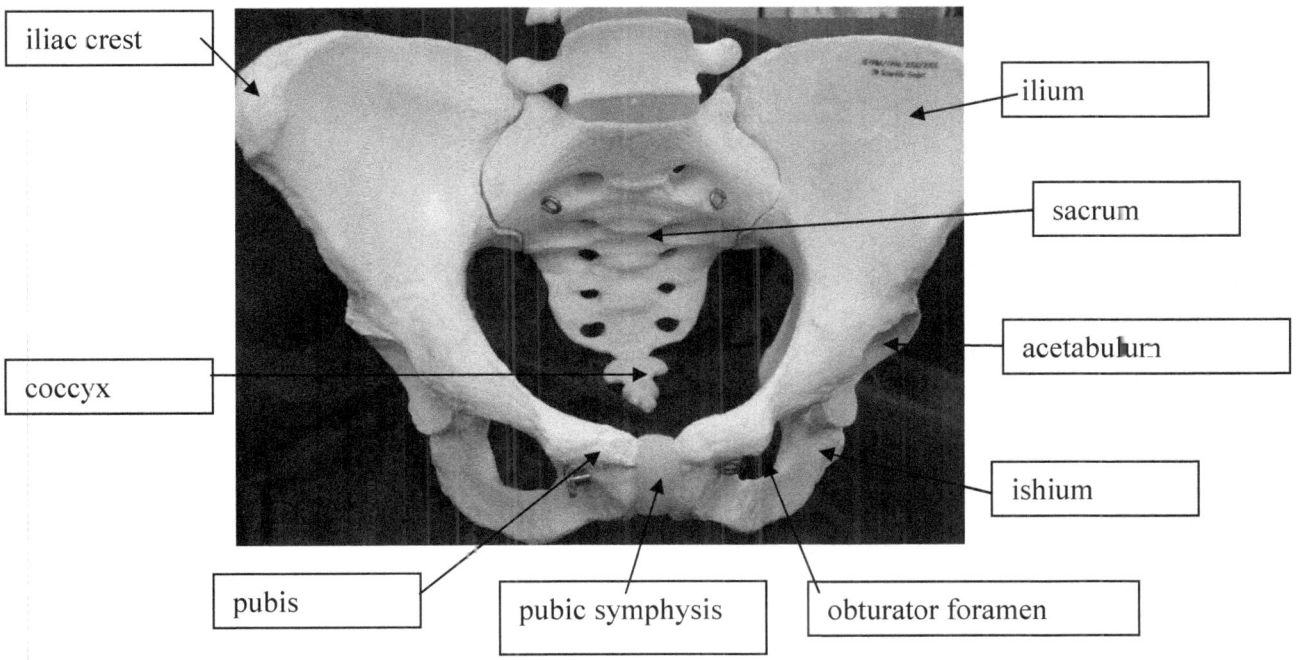

- iliac crest
- coccyx
- pubis
- ilium
- sacrum
- acetabulum
- ishium
- pubic symphysis
- obturator foramen

Pelvic Girdle (posterior aspect)

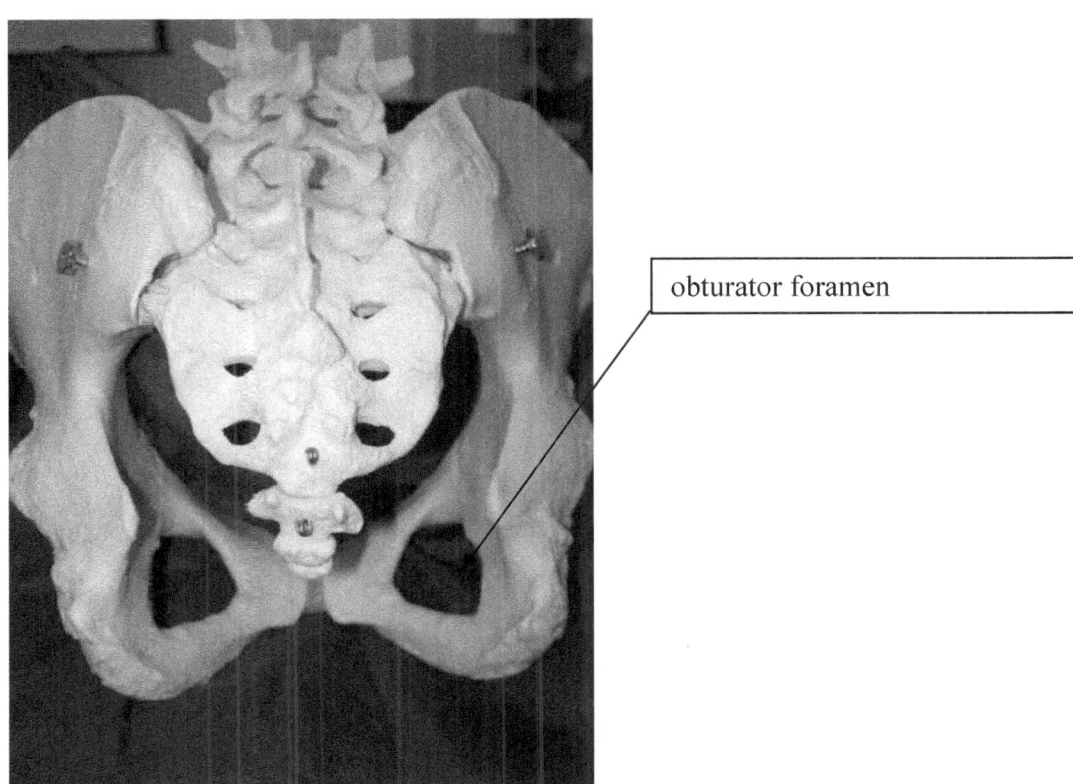

- obturator foramen

Name _____ Class Meeting Day ____ Class Time _____

Pelvic Girdle (lateral aspect)

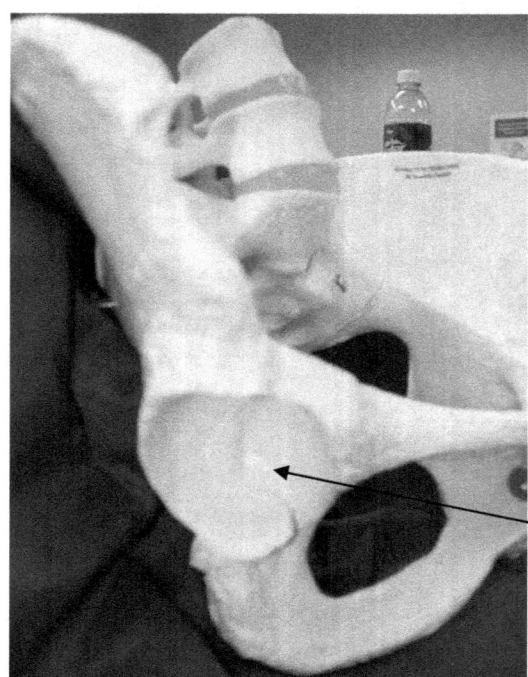

acetabulum

Human Anatomy and Physiology Laboratory Exercises—Level 1

Using Crime-Scene Investigative Approaches

Stephanie A. Lanoue

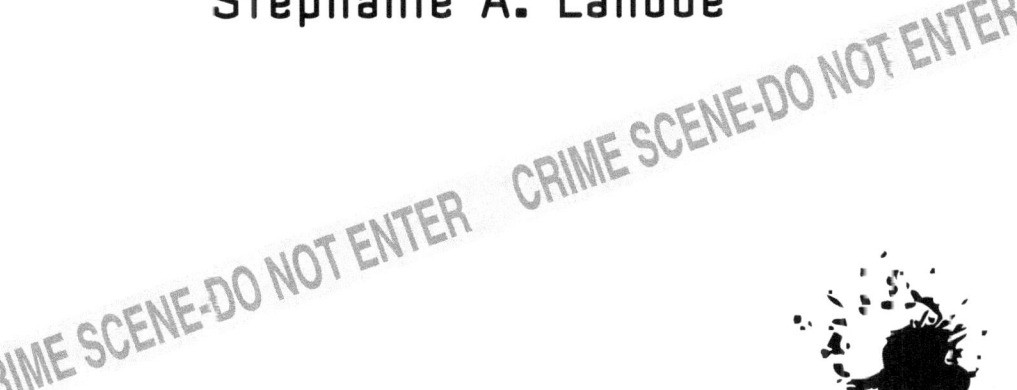

Cover image credits:

Background © echo3005, 2009. Used under License from Shutterstock, Inc.
Thumb print © mehmetsait, 2009. Used under License from Shutterstock, Inc.
Outline of body © Linda Bucklin, 2009. Used under License from Shutterstock, Inc.
Hand print © Vladmir Zivkovic, 2009. Used under License from Shutterstock, Inc.
Face © dundanim, 2009. Used under License from Shutterstock, Inc.
Microscope ©Svon Hoppe, 2009. Used under License from Shutterstock, Inc.
Hand holding vial of blood © Dusan Zidar, 2009. Used under License from Shutterstock, Inc.
Hand holding flask © Jon Kroninger, 2009. Used under License from Shutterstock, Inc.
Blood splatter © Rae, 2009. Used under License from Shutterstock, Inc.

Kendall Hunt
publishing company

www.kendallhunt.com
Send all inquiries to:
4050 Westmark Drive
Dubuque, IA 52004-1840

Copyright © 2009 by Kendall Hunt Publishing Company

ISBN 978-0-7575-6625-7

All rights reserved. No part of this publication may be reproduced,
stored in a retrieval system, or transmitted, in any form or by any means,
electronic, mechanical, photocopying, recording, or otherwise,
without the prior written permission of the copyright owner.

Printed in the United States of America
10 9 8 7 6 5 4 3

Dedication

This book is dedicated to Richard, my husband of 21+ years, who encouraged me to write this book despite my busy schedule and for every student who ever dreamed of a non-boring Anatomy & Physiology course.

 # CONTENTS

Preface   vii

About the Author   ix

Laboratory Exercise 1   Crime—Scene Investigation and Orientation to the Human Body   1

Laboratory Exercise 2   The Microscope as a Forensics Tool   9

Laboratory Exercise 3   The Case of the Missing Computer at LIT   17

Laboratory Exercise 4   DNA Extraction   25

Laboratory Exercise 5   The Case of the Missing Histology Slides   31

Laboratory Exercise 6   A Closer Look at Skin   41

Laboratory Exercise 7   Whose Hair Is This? Using Hair to Determine Species and Other Quirky Facts   49

Laboratory Exercise 8   Bone Anatomy: Gaining Familiarity with Bone Markings and the General Skeleton   59

Laboratory Exercise 9   Bone Forensics: What Are The Bones Telling you?   71

Laboratory Exercise 10   Anatomy and Characteristics of Human Muscle   79

Laboratory Exercise 11   The Nervous System Versus the Case of the Distracted Driver   95

Laboratory Exercise 12   That's Using Your Senses! The Case of the Missing Test and the Student's Ear   105

 # PREFACE

I've always thought that it is easier to learn something when we are interested in it or when it is very relevant to our lives. Perhaps you have seen popular television shows like *CSI-Miami, CSI-New York, Eleventh Hour, Bones,* or *Forensic Files*? If you're like me, you may enjoy trying to solve crime scenes and guessing who committed the crime. Solving crime scenes and using forensic evidence is based on solid science. Many times, this science involves knowledge of the human body such as anatomy and physiology.

It is my hope that combining CSI (crime scene investigation) with anatomy and physiology will not only make the laboratory portion of this course more interesting but it will also add relevance and meaning for students. I designed the laboratory exercises to be memorable and for college-level students to actually have fun while learning anatomical parts, making predictions, role playing, tackling problem-solving scenarios, and working collaboratively in small forensic teams to achieve learning outcomes and objectives.

Further, each laboratory exercise is built upon a solid, well-researched model for teaching science called the 5E model. There are literally five "Es" comprising the model—each word begins with the letter "e." The first "E" represents the "Engage" portion of the lesson. This portion of the exercise is designed to grab the learner's attention by providing an opportunity to make a prediction, reflect on past experiences, think about the solution to a problem, or simply gain background information needed to solve the "case." The second "E" stands for "Exploration." Exploration is the very nature of science! It allows students to work "hands-on" and to get involved. The third "E" is for the "Explanation" phase. Students are encouraged to develop solid explanations or back up an original hunch or prediction. The fourth and fifth "Es" stand for "Extend" (Extension) and "Evaluate" (Evaluation) respectively. Extension challenges the student to take things a step further, investigate other possibilities, or conduct research into something that he or she is interested in. Last, evaluation provides an opportunity to summarize what has been learned in the student's own words. The 5E model helps to ensure that students do more than simply memorize body parts—it promotes higher order thinking skills by involving them in analyzing situations, solving problems, and drawing conclusions.

I hope you'll put on your CSI thinking cap and join me for some new adventures in learning. Enjoy!

ABOUT THE AUTHOR

Stephanie A. Lanoue is interested in bringing biology to life for students. She believes that science is relevant to student's lives and that anything that establishes that relevance such as a television show or a real-life experience should be utilized in the classroom setting. She also believes that students who are genuinely interested will in turn ask questions, explore, and become much more engaged in the learning process. Her focus is on problem-solving methodologies that get adult learners to use higher order thinking skills rather than just low-level thinking skills like simple recall and memorization.

Stephanie has a lifetime of experience in both education and science. She began her adventure in science teaching at the young age of 21 in the public schools of Texas as a high school biology teacher. After numerous years at that level, she decided to pursue teaching at the college level. She currently teaches fulltime Anatomy & Physiology for the Lamar Institute of Technology in Beaumont, Texas. Additionally, Mrs. Lanoue has served as a science consultant on the state-wide level. She has been invited to speak at national conferences and has received numerous awards including being chosen as 1 of 8 finalists on the national level for Outstanding Technical Teaching.

When she is not teaching, Stephanie actively pursues creative endeavors. She is an accomplished wildlife artist/ illustrator, writer and musician. She is happily married to her husband, Richard, of 21+ years and they share their household with a gregarious rough-coat Jack Russell terrier named Annie.

LABORATORY EXERCISE 1
Crime-Scene Investigation and Orientation to the Human Body

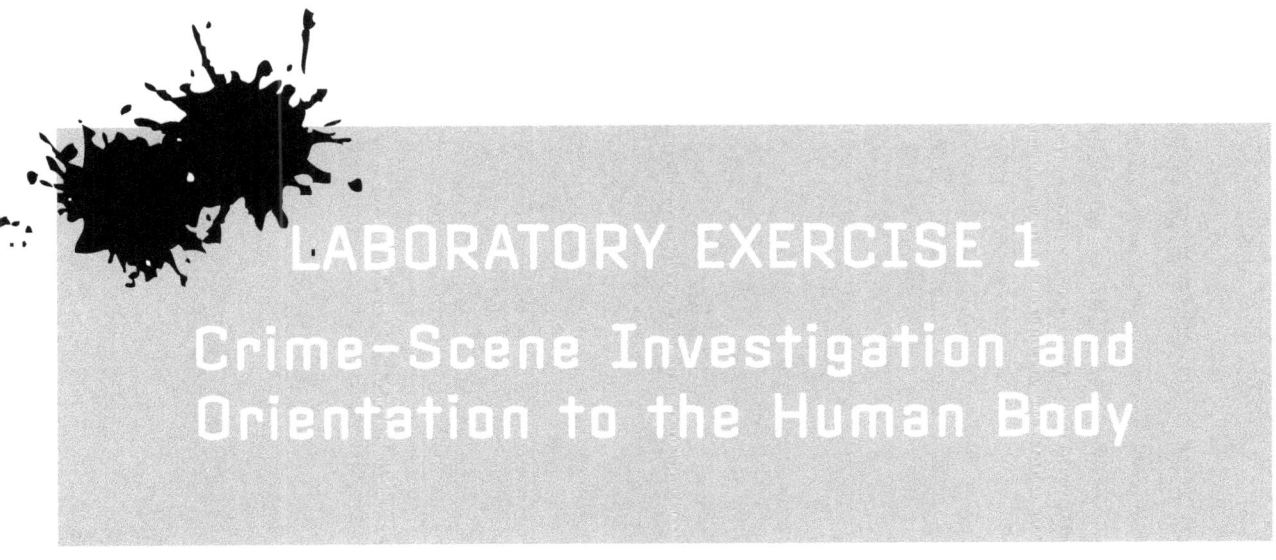

Learning Objective(s)

The student will:
- Describe body orientation at various crime scenes using correct anatomical terms.
- Demonstrate the planes of the body.

MATERIALS: human model (half-torso or full), 3 gummy bears, 3 microscope slides

Your Name: _____   Today's Date: _____

Class: _____   Class Time: (Day) _____ (Time) _____

I. **Instructions:** Working in small teams of 2–3 students, familiarize yourselves with the following background information addressing crime-scene investigative policies.

HANDLING HOMICIDE OR SUSPICIOUS DEATH SCENES

The first officer to arrive at a homicide or suspicious death scene notifies his/her immediate supervisor, who then notifies the Homicide Unit and the on-duty Watch Commander by telephone, when possible. Homicide Unit investigators notify the Crime Lab Unit. The first officer at the scene immediately takes the following measures:

- Requests assistance in securing the scene.
- Secures the area where the crime was committed and where the body was found. For an indoor scene, seal off all points of entry and exits. For an outdoor scene, use crime scene tape to immediately rope off the area, securing the largest area possible. Efforts should be made to protect evidence from the elements. Be aware of the possibility of multiple scenes.
- Do not allow any unauthorized persons to enter or leave the scene. Identify any unauthorized persons attempting to enter the scene and relay this information to the on-scene investigator(s). No one shall be permitted to approach the scene until investigators arrive.
- Do not allow anything to be touched or moved at the scene. Do not move a weapon unless it presents an immediate danger. Do not cover the body with anything or allow anyone else to cover the body. If possible, and without disturbing evidence, create a secure environment around the body so that it is not in plain view of the public until the arrival of Homicide or Identification Division personnel.
- Upon arrival at the scene, Homicide investigators or Crime Lab personnel are responsible for placing a barrier around the body. The decision of whether to cover a body is made by a Homicide investigator or the Commander of the Homicide Unit.
- If covering the body is not possible, the Commander at the scene will use other methods that might be available, i.e., a vehicle, bus, etc. placed to obstruct viewing by the public.
- If the decision to cover a body has been made, the deceased will be covered with a sterile sheet provided by the Medical Examiner's Office. All Homicide cars have a supply of pre-packaged sheets. The package in which the sheet was enclosed will be property inventoried. The sheet will be taken by Medical Examiner's Office staff.

Your Name: _____ Today's Date: _____

Class: _____ Class Time: (Day) _____ (Time) _____

- If anything is touched or moved, inform investigators and document it. Do not remove anything from the scene unless instructed by investigators.
- Identify and isolate <u>all</u> witnesses as soon as possible. Identify witnesses by name, date of birth, home address, telephone number, business address, business telephone, and business hours. All witnesses will be transported to the Homicide Unit at the direction of the investigator(s) and/or Watch Commander. If a witness refuses to go to the Homicide Unit, inform the investigator(s) or Watch Commander.
- Eliminate sightseeing at the scene, including police. Do not discuss the incident; direct all requests for information to the on-duty Watch Commander.
- Upon arrival, investigators are in charge of the scene. The supervisor reports to the investigator(s) in charge at the scene and provides assistance as needed. When released from the scene and based on staffing levels, the supervisor directs officers to complete required reports. All police personnel present at the scene make a statement unless instructed otherwise by investigators.

Photographs at death/homicide scenes are taken only by Crime Lab personnel or by persons authorized by the Homicide Unit investigators.

NOTIFICATION OF THE MEDICAL EXAMINER

In the event of a homicide or suspicious death, an on-scene Homicide investigator must notify the Medical Examiner's office as soon as possible.

In all other deaths, officers notify the Medical Examiner's office by telephone, even if a qualified physician is present. Personnel from the Medical Examiner's office determine whether they will investigate the scene or release the body to a funeral home.

In summary, crime scene investigation is diligent work. In a homicide case, it is particularly important to accurately document the exact position of wounds, weapons, etc. The description should be provided in exact anatomical terms.

II. Working in a small group, help complete the crime scene descriptions by selecting an anatomical term that appropriately completes the crime scene. Then, demonstrate the locations/anatomical terms on an anatomical model with others in your group. You will be responsible for knowing the following anatomical terms in respect to an anatomical model.

Words to choose from: *superior, inferior, ventral (anterior), dorsal (posterior), medial, lateral, intermediate, proximal, distal, superficial* or *deep*.

Your Name: _____   Today's Date: _____

Class: _____   Class Time: (Day) _____ (Time) _____

The words may be used more than once. And, depending on perspective, more than one term may apply to the situation (in that case—pick the term that you feel is most appropriate). Don't forget to demonstrate the terms/locations on a model!

Crime Scene Findings (written in common terms)	Complete the Description Using the Correct Anatomical Terms
a. The cut was shallow and did not penetrate any muscle.	The cut is _____ to the muscles.
b. A stab wound was also identified on the right shin.	The right shin is _____ to the right knee and has a stab wound.
c. A second victim was found. A gunshot entered the front of the body and penetrated the sternum (breast bone).	The gunshot entry point occurred to the _____ side of the body. The wound is also _____ to the sternum.
d. A cadaver has a knife wound in the left thigh approximately 5 inches above the knee.	The knife wound is _____ to the left knee. (NOTE: DO NOT USE the term superior)
e. The bullet exited the victim's body out the spine.	The bullet exited the victim's spine, which is the _____ side of the body.
f. Apparently, what killed the victim was a bullet to the heart.	The heart is _____ to the arms.
g. In yet a third case, a victim was found with both arms missing but chest (thorax) in tact.	The missing arms are _____ to the victim's chest.
h. The victim was found face down.	The victim was found _____ side up.
i. The victim's nose was broken.	The nose is _____ to the scalp and the chin.
j. The accident ruptured the spleen.	The spleen is _____ to the skin.
k. Falling from the roof fractured the victim's elbow and dislocated his shoulder.	The shoulder is _____ to the elbow.
l. Fortunately, the bullet only grazed the shoulder, striking the elbow with full impact.	The elbow is _____ to the shoulder.

Your Name: _____   Today's Date: _____

Class: _____   Class Time: (Day) _____ (Time) _____

III. Locate the following areas on an anatomical model:

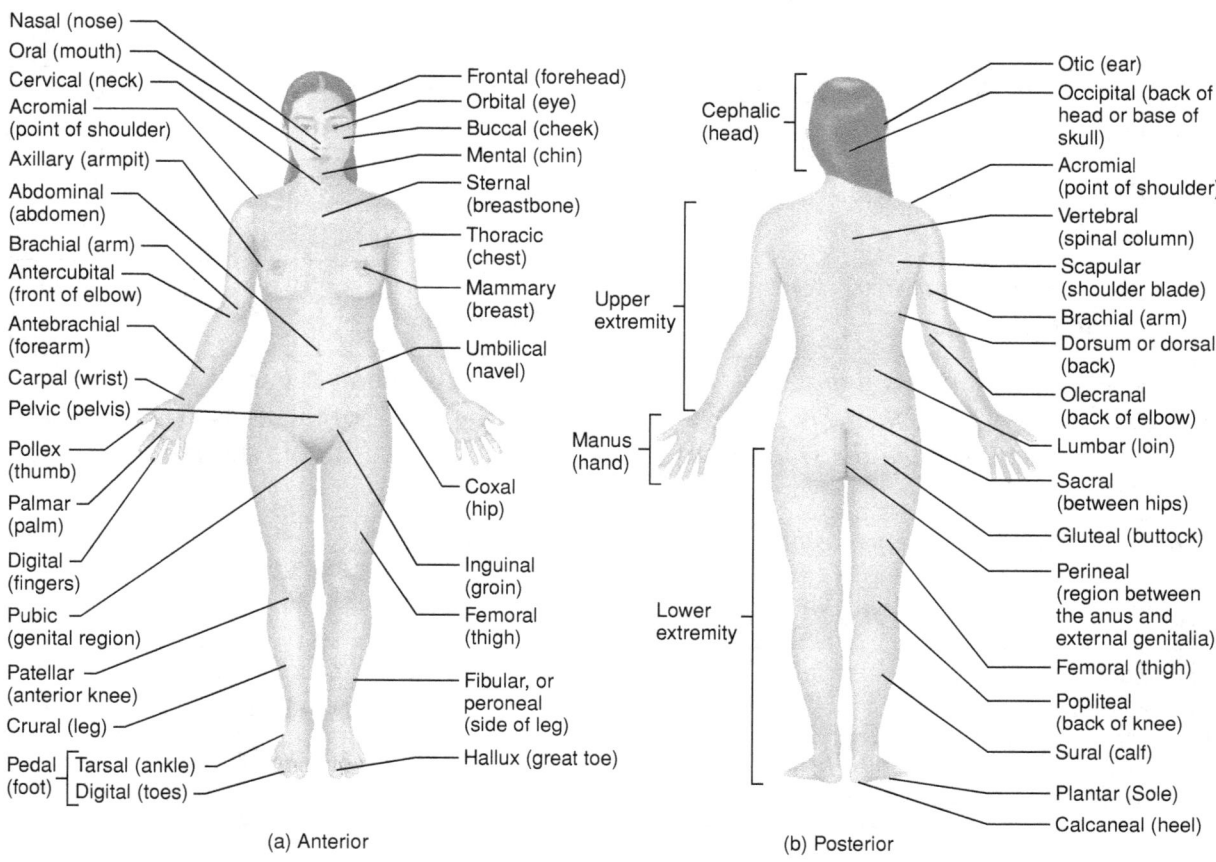

Follow-up Questions:

a. Closer examination revealed a broken pollex. This is a broken _____.

b. If you cut your sural area, you have an injury to the _____.

c. Plantar warts are known to grow only on the _____ of the _____.

d. Phlebotomists typically draw blood from the veins located in the antecubital region also known as the _____.

Your Name: _____   Today's Date: _____

Class: _____   Class Time: (Day) _____ (Time) _____

IV. Extension

A. Using gummy bears and microscope slides, demonstrate to your group the following sagittal planes. Use a microscope slide to bisect each gummy bear:

Gummy Bear 1: Frontal section through the torso. Sketch your results:

Gummy Bear 2: Transverse section through the torso. Sketch your results:

Gummy Bear 3: Median or midsagittal section through the torso. Sketch your results:

B. Concluding Questions:

1. Which of these planes would separate the neck from the knees?

2. Which of these planes would separate the navel from the spine?

Your Name: _____   Today's Date: _____

Class: _____   Class Time: (Day) _____ (Time) _____

   3. Which plane would separate the left leg from the right leg?

   4. If an individual is shot and a bullet enters the thoracic cavity through the sternum, what organs might be in danger of injury?

V. Conclusion

   A. In your own words, what did you learn from today's laboratory exercise with regard to anatomical terms and crime scene descriptions?

   B. How might the planes of the body apply to medical testing such as CT scans and MRIs?

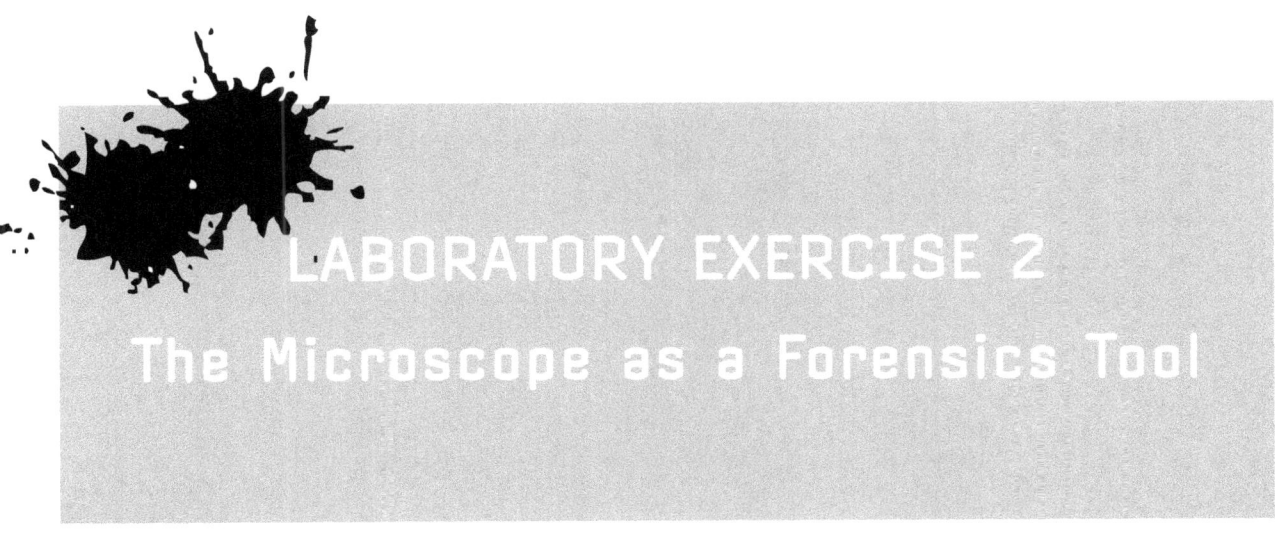

LABORATORY EXERCISE 2
The Microscope as a Forensics Tool

Learning Objective(s)

Working in teams of two, the students will:

- Learn the parts of a binocular microscope and demonstrate their functions with regard to focusing, light manipulation, and maneuvering the mechanical stage.
- View biological evidence from a crime scene (a prepared blood slide) using a full range of objective lenses (i.e., 4×, 10×, 40×, and 100×) and create realistic sketches of what they see.

MATERIALS: binocular microscope, prepared blood slide, no. 2 pencil, colored sketching pencils

Your Name: _____   Today's Date: _____

Class: _____   Class Time: (Day) _____ (Time) _____

I. Introduction

A. Perhaps you have seen a forensic crime scene technician using a microscope on a popular investigative television show? Perhaps you have used one yourself? The microscope is a powerful forensics tool and allows evidence to be viewed at many times normal vision.

(1) Describe any previous experience you have using a microscope.

(2) How would you rate your confidence level using a microscope before beginning this laboratory exercise? (CIRCLE ONE)

```
      0                  1                2                  3
(Have not used one)  (Not Confident)  (Confident)   (Very Confident)
```

B. Obtain a binocular microscope. Using the labeled photograph (Figure 1), find each part on your binocular microscope. Each partner takes turns finding the parts below. Complete the checklist as you go along.

- __ A. Eyepiece
- __ B. Dioptric adjustment ring
- __ C. Objectives (lenses)
- __ D. Working stage
- __ E. Mechanical slide holder (clip)
- __ F. Light Diffuser
- __ G. Condenser focus knob
- __ H. Iris diaphragm (Note: looks like a silver slider)
- __ I. Lighting Brightness Wheel
- __ J. Power Switch
- __ K. Coarse focus adjustment
- __ L. Fine focus adjustment
- __ M. Mechanical stage adjustment

Your Name: _____   Today's Date: _____

Class: _____   Class Time: (Day) _____ (Time) _____

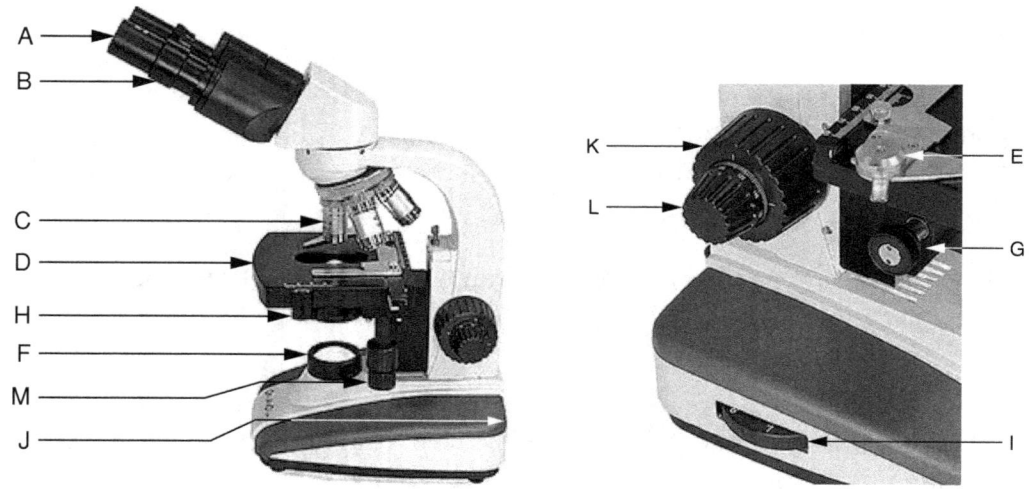

Figure 1

C. Now label the "blank" microscope illustration below (Figure 2) with the proper part name. Challenge yourself and try to label without viewing your notes.

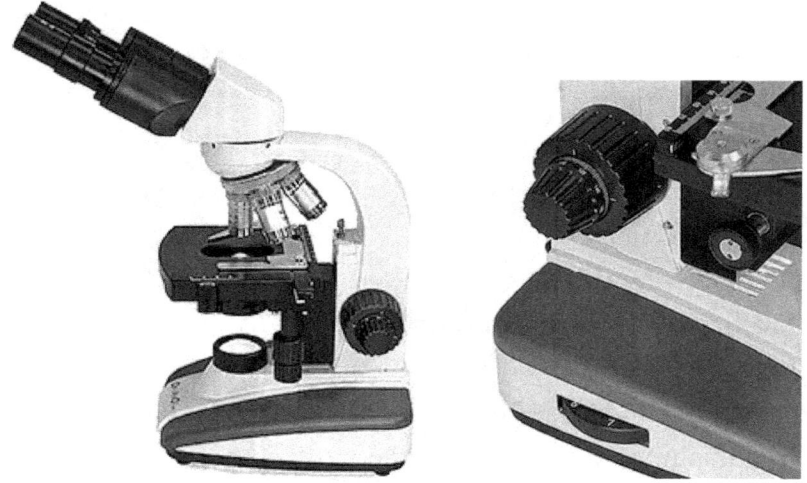

Figure 2

II. Predict

What do you think blood will look like under the microscope? (describe your hypothesis) _____

_____

_____

_____

Your Name: _____   Today's Date: _____

Class: _____   Class Time: (Day) _____ (Time) _____

III. Using the Microscope to View a Specimen

Obtain a prepared blood slide. View the slide using the microscope.

Use the checklist below to help:

___ a. Be sure microscope is uncovered and plugged in.

___ b. Be sure eye covers (small plastic protective covers) are removed from eyepieces. **NOTE:** *Eyepieces can be adjusted to your eye width.*

___ c. Turn on microscope.

___ d. Place slide on stage. Place it in the mechanical stage holder. **NOTE:** *The long clip is movable—make sure your slide is "seated" within the mechanical stage holder.*

___ e. Starting with the lowest power objective lens (4×), view the slide. **NOTE:** *Always start with 4× when first viewing a slide.*

___ f. Using the mechanical stage adjustment knob, move the slide into the field of light (light is coming from the diffuser and the amount of light may be controlled by the light brightness wheel). **NOTE:** *It may help if you "eye" the slide at the same time with your own set of eyes; if the slide appears to have something on it, line that up with the light coming up.*

Follow-up Questions (Describe in relation to your body)

1. When you turn the top (larger) mechanical stage adjustment knob clockwise, which way does the slide (or image) move? _____

2. When you turn the bottom (smaller) mechanical stage adjustment knob clockwise, which way does the slide (or image) move? _____

3. When you turn the top (larger) mechanical stage adjustment knob <u>counterclockwise</u>, which way does the slide (or image) move? _____

4. When you turn the bottom (smaller) mechanical stage adjustment knob <u>counterclockwise</u>, which way does the slide (or image) move? _____

___ g. Using the <u>coarse focus adjustment</u>, bring the slide into focus.

Laboratory Exercise 2

Your Name: _____   Today's Date: _____

Class: _____   Class Time: (Day) _____ (Time) _____

___ h. Sketch what you see: (sketch with careful attention to detail)

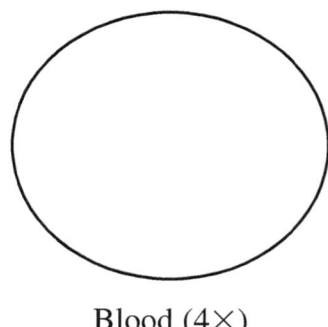

Blood (4×)

IV. Using Higher Objectives and Fine Focus Adjustment

A. Next, you will view blood on a higher objective (10×). This will enable you to see the blood at a higher magnification.

B. Rotate the objective lens to 10×. Be sure to snap the lens in place.

C. Focus the slide using <u>coarse adjustment</u>.

D. Draw what you see:

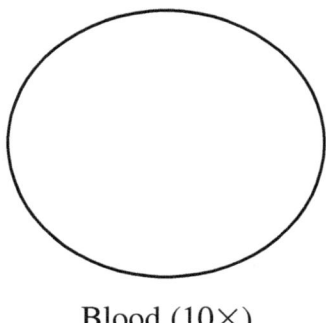

Blood (10×)

E. Next, you will view the blood at 40× power.

F. Rotate the objective lens to 40× carefully. Be sure it clears the stage. It is ok to use coarse focus on 40×, but turn the knob VERY slowly.

Your Name: _____   Today's Date: _____

Class: _____   Class Time: (Day) _____ (Time) _____

G. Draw the blood on 40×.

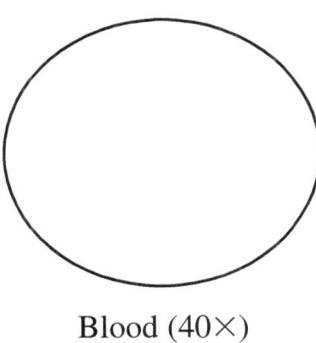

Blood (40×)

H. Next, you will view the blood on 100×. You will need lots of light to see the blood on 100× so adjust all light sources to "bright." Slowly focus the slide using the <u>FINE adjustment knob</u> only. **NOTE:** *When you are using 100× or higher, ALWAYS use fine adjustment (never coarse adjustment—you could break the lens or slide or both).*

I. Read the following about the three objective lenses and their focusing ability:

<u>The objective lenses in your microscope are parfocal.</u> *Parfocal* means that the <u>microscope objectives</u> stay in focus when magnification is changed; i.e., if the <u>microscope</u> is switched from a lower power objective (e.g., 10×) to a higher power objective (e.g., 40×), or vice versa, the object stays in focus

Because the microscope lenses are parfocal, you should barely need to use the fine focus adjustment to bring the blood into focus.

J. Draw the blood on 100×.

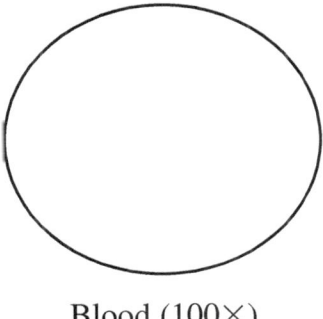

Blood (100×)

Your Name: _____ Today's Date: _____

Class: _____ Class Time: (Day) _____ (Time) _____

> **NOTE:** *The 100× objective lens is also called the "oil immersion" lens. Oil immersion is a technique typically used with fresh samples (check tissue, etc.) A special drop of oil is placed between the tissue and the objective lens. No cover slip is used.*

K. At this time, if you have not already, try some of the microscope's features and adjustments such as the light brightness wheel, dioptric adjustment ring, or iris diaphragm.

V. Analysis of Microscope Use

A. In your own words, explain the concept of parfocal lenses.

B. What focus adjustment knob must be used with the 100× lens and why?

C. How would you rate your confidence level using a microscope after finishing this laboratory exercise? (CIRCLE ONE)

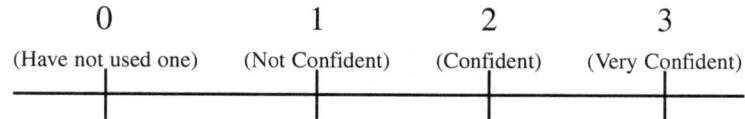

VI. Conclusion

In your own words, summarize what you learned today about using the microscope.

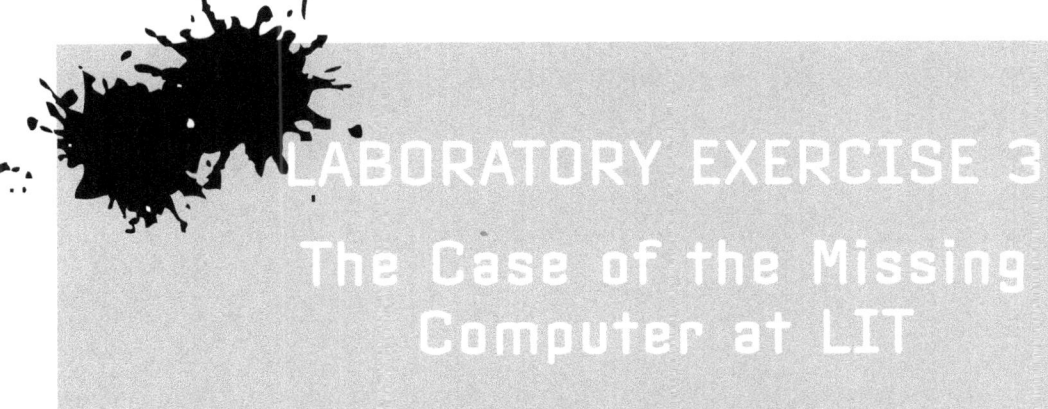

LABORATORY EXERCISE 3
The Case of the Missing Computer at LIT

Learning Objective(s)

The student will be able to:

♦ Use microscopy techniques to examine forensic evidence to solve a "staged" crime scene.

MATERIALS: microscope, evidence A (box of microscope slides that includes a red liquid-like substance, newspaper fragment, white crystal-like substance, dust, tiny portion of an insect, and a 1 mm cork-like solid fragment), evidence B (box of microscope slides that includes a tiny portion of an insect, computer chip, red liquid-like substance, white crystal-like substance, and a cork-like solid fragment.)

Your Name: _____   Today's Date: _____

Class: _____   Class Time: (Day) _____ (Time) _____

I. Introduction

Divide into small groups of 3–4 people. In this lab, you will investigate a crime scene report and use a microscope to study evidence gathered from the crime scene. Each person on the team will assume one of the following roles:

Forensic Team Responsibilities

Chief of Forensics This individual is in charge of the entire investigation. Additionally, he or she is responsible for the accuracy of the final Forensic Report and for briefing the team on the proper use of the microscope and microscopic techniques. This person should be the best at using a microscope (i.e., help your team focus the microscope, scan slides, keep things moving, etc.)

Investigative Reporter This individual's primary responsibility is to help prepare the written Forensic Report after reviewing the evidence, preliminary police report, and collaborating with other Forensic Team members.

Evidence Recorder This person is responsible for obtaining and returning the evidence (making sure it is handled properly and put back properly). Additionally, the Evidence Recorder will prepare the evidence for viewing with a compound light microscope. All evidence must be properly observed and diagramed/logged onto a Forensic Laboratory Evidence Sheet.

Forensic Artist This person will sketch the evidence during microscopic study. Thorough documentation includes a detailed, scientific sketch of each piece of evidence collected. These sketches should be included in the Evidence portion of the Forensic Report.

II. Familiarize Yourselves with the Crime Scene Reports:

THE PRELIMINARY POLICE REPORT—CRIME SCENE A

Date: 8/23/08

At approximately 7:00 am, a theft occurred at the LIT Multi-Purpose building located at 544 E. Lavaca, Beaumont, Texas 77705. Police arrived on the scene at 7:30 am after receiving a call from a despondent staff member who discovered the situation upon arrival at work. The staff member reported finding the faculty break room (Room 225) in disarray and the "community" computer, typically used as a common workstation by faculty/staff, missing.

Your Name: _____ Today's Date: _____

Class: _____ Class Time: (Day) _____ (Time) _____

Investigators found what appeared to be drops of a red substance on the floor of the break room. They also collected samples of other items from the surfaces in the break room as circumstantial evidence. Those items included a torn paper fragment, white crystal-like substance, dust, a tiny portion of an insect, and a cork-like solid fragment measuring approximately 1 mm in diameter. Investigators noted that a salt shaker had been turned over and salt was apparently spilled on the table and floor. It looked as if a partial shoe print was evident in the spilled salt on the floor although it was rather nondescript.

Following a thorough search of the grounds and parking areas, investigators found a car without a license plate or LIT parking tag parked to the rear of the north lot. There were drops of a red substance found on the ground outside the car. There was also a sponge-like fragment on the ground near the red substance. The car appeared abandoned. Police set up a watch on the vehicle from a remote location.

FOLLOW-UP REPORT—CRIME SCENE B

At approximately 11:45 am, a suspicious-looking individual wearing a hooded black jacket and black slacks approached the car carrying a large cardboard box. The face was not visible. The person attempted to unlock the car and load the box into the car. Police approached the individual and startled him/her. The individual dropped the box and ran. The individual had a hard time running away . . . apparently the sole of the shoe was damaged in some manner. The individual kicked off shoes and fled. Police were able to apprehend the individual and take him/her into custody.

Investigators discovered that the box held a flat-panel computer monitor only. There was also a portion of a dead insect in the box. On the ground beside the box lay a computer chip that was also gathered for evidence. The box was smeared with a red substance.

When the shoes were examined, the sole of the left shoe had a white, crystal-like substance adhering to it. The right shoe had a small chunk missing from it . . . the inner soles appeared to be cork.

III. Preparing the Forensics Report

FORENSIC REPORT GUIDE

In your forensic report you will outline the crime scene and evidence as you understand it. You will present this as a report to the Chief of Police (the instructor). The Investigative Reporter's primary responsibility is for the processing and submittal of this report. Be sure to follow proper reporting format.

Human Anatomy and Physiology

Your Name: _____ Today's Date: _____

Class: _____ Class Time: (Day) _____ (Time) _____

Complete your report as follows:

Investigation of Incident at:
Location:
Address
City, State, Zip:
Room:
Date:

List the names of your team members by role:

<u>Chief of Forensics</u>–
<u>Evidence Recorder</u>–
<u>Investigative Reporter</u>–
<u>Forensic Artist</u>–

Prepare the report with the following sections:

A. REPORT INTRODUCTION:

What do you know about the scene that is based on the facts—who, what, when, where, why. WRITE JUST THE FACTS. What do you know so far?

B. HYPOTHESIS:

Consider each piece of evidence collected and what you believe it to be at this point in the investigation. You may change your mind later.

Laboratory Exercise 3

Your Name: _____   Today's Date: _____

Class: _____   Class Time: (Day) _____ (Time) _____

Substance Collected	Hypothetical Possibilities
Red liquid substance	
White crystal-like substance	
Tiny portion of an insect	
Solid, cork-like fragment	
Torn paper fragment	
Computer chip	

C. POSSIBLE MOTIVE

Based on your teams analysis of the evidence → What conclusions are you able to reach thus far? Why do you feel this way? Do you believe there is a motive?

D. EVIDENCE SKETCHES

Support your hypothesis (what PROOF do you have?). Include ALL of your evidence in this section of your report. What evidence did you receive? What does the evidence show/prove? What conclusion(s) can be made about what actually occurred? Use the evidence to support your conclusion(s).

Crime Scene A (sketches from Evidence Box A)

Your Name: _____ Today's Date: _____

Class: _____ Class Time: (Day) _____ (Time) _____

Crime Scene B (sketches from Evidence Box B)

E. FINAL CONCLUSION/VERDICT

Does the evidence at crime scene B (parking lot) place the individual back at crime scene A (faculty break room)? What conclusion does your team support? Do you believe there is sufficient evidence to make a conviction? Why or why not? Be sure to state which evidence pieces led you to this conclusion.

Examples: "The evidence _____, _____, and _____ leads us to conclude . . . or "It appears that . . . so and so is/is not guilty because of the following . . ." Write your team's conclusion below.

Laboratory Exercise 3

LABORATORY EXERCISE 4
DNA Extraction

Learning Objective(s)

The student will be able to:

- ◆ Recognize that DNA is found in cells.
- ◆ Extract DNA and explain the steps needed to isolate DNA from a cell.
- ◆ Describe the structure of DNA

MATERIALS: 1 frozen strawberry (students can easily work in teams of two with one strawberry for every two students). 1 self-sealing sandwich bag, 1 15 ml test tube or 5 oz paper bathroom cup, 1 clear glass or plastic test tube, 1 large gauze square, rubber band, 10 ml extraction solution (see recipe below), 1 bamboo skewer or glass stirring rod

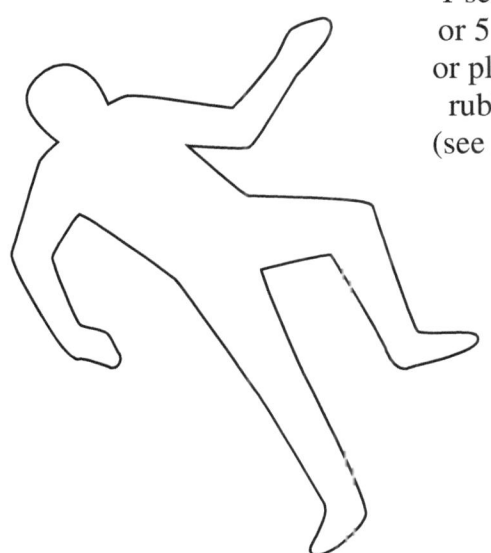

Your Name: _____ Today's Date: _____

Class: _____ Class Time: (Day) _____ (Time) _____

Extraction solution recipe:

- 450 ml distilled water
- 10 g table salt
- 50 ml liquid ishwashing detergent

For the whole class to share:

- Clean food bowl
- 90% ice cold rubbing alcohol or ethanol
- Eye droppers for dispensing solutions

I. Introduction

Strawberries are used in this activity because they are octaploid, meaning they have eight copies of every gene rather than the usual two, thus providing large quantities of DNA to extract. Naturally, strawberries are also relatively inexpensive and readily available. Other great sources of DNA to experiment with include kiwis and calf liver.

The DNA molecule is typically a very long, thin strand that is coiled tightly in the nucleus of a cell. It is typically not visible to the naked eye and requires a very high-powered microscope to see it inside the nucleus of a cell. The DNA found in each human cell is almost 2 meters long. If all the DNA in a human adult (that's 100 trillion cells) were laid end to end, the DNA would stretch 113 billion miles. That would take you to the sun and back 610 times. In this activity, no microscopes are needed! You will uncoil the DNA molecules and free them from the nucleus. The DNA strands will tangle together into a thick rope-like molecule, visible without magnification.

A. What do you already know about DNA? (i.e., where it is, why is it unique, etc.?)

B. What do anticipate it will look like when extracted from the cell? (i.e., color, texture, etc.?)

Laboratory Exercise 4

Your Name: _____ Today's Date: _____

Class: _____ Class Time: (Day) _____ (Time) _____

C. Take a moment and view a stained/prepared human cell such as a check cell under a microscope at 100× magnification. Study the nucleus. Make a sketch of what you see. Label the nucleus, cytosol, and cell membrane.

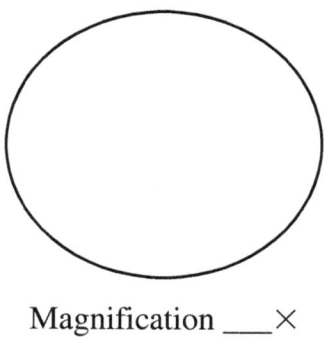

Magnification ___×

1. Describe the DNA in the nucleus.

2. Can you see the double-helix molecular structure? Why or why not?

II. Procedure Overview

The process itself is fairly straightforward. First, the cell walls are broken open by smashing the strawberries in the sandwich bag. Next, detergent is used to dissolve the cell and nuclear membranes. The membranes are made of lipids (fat) and the detergent will cut through the membrane just like it cuts through grease on a dirty plate when washing dishes. Some salt is present in the detergent solution in order to match the osmolarity of the cells.

Now you have a big mixture of smashed cell walls, dissolved membranes, loose DNA and random other cell parts. This mixture is filtered through paper towels or gauze. Finally, you take advantage of the fact that DNA is soluble in water but not in alcohol. In fact, alcohol makes DNA clump together. Thus a layer of alcohol is laid on top of the filtrate. Any DNA that makes contact with the alcohol will clump together, pulling the rest of the DNA strand along with it. Watch for tiny white strands of DNA bubbling their way up from the red strawberry extract.

Your Name: _____   Today's Date: _____

Class: _____   Class Time: (Day) _____ (Time) _____

The DNA may be collected by twirling a bamboo skewer or glass stirring rod in the solution. The DNA strands will spool themselves around the skewer and can be pulled out of the solution.

III. Procedure

1. Pass out plastic sandwich bags and strawberries.

2. Students should put the strawberry in the sandwich bag, squeeze out most of the air, and seal the bag. The strawberry can then be crushed into juice and pulp. Try to squish all of the chunks into an even, smooth puree. Warn students not to pound the strawberry on the table or risk the bag bursting.

3. Next, open the bag and add 10 ml of extraction solution (approximately 10 eyedroppers full). Seal the bag again and gently mix the strawberry juice with the extraction solution. Warn students not to mix too vigorously or it will generate a lot of bubbles and can't be filtered effectively. Use a gentle tilting back and forth motion while lightly squeezing the bag.

4. Set up a filtration system. Use a rubber band or hold a large gauze square over a paper cup.

5. Carefully pour the extract onto the gauze. Allow the juice to filter through the gauze into the container below. Let it drip for 3–5 minutes. Do not squeeze the gauze or rush the process.

6. The gauze can be put inside the sandwich bags and thrown away.

7. Carefully transfer liquid from the 15 ml tube or cup into the clear test tube until the test tube is about a third full.

8. Slowly add 3 ml (3 eye droppers full) of ice-cold alcohol to the test tube. The alcohol should be added so that it trickles down the side of the tube before pooling on top of the strawberry extract. You should end up with a red bottom layer and a clear top layer.

9. After 2–3 minutes, a skewer or stirring rod can be inserted into the tube and gently swirled around. This will spool the DNA around the stick. The DNA can be pulled out of the tube and viewed. Students may safely touch the DNA, although the DNA should NOT be tasted under any circumstances.

Your Name: _____   Today's Date: _____

Class: _____   Class Time: (Day) _____ (Time) _____

IV. Conclusion

A. Follow-up Questions

1. What is the purpose of mashing the strawberry in the sandwich bag?

2. Which procedures allowed for the release of DNA from the cells?

3. How are oil and water like DNA and alcohol?

4. Why does DNA rise to the top after adding alcohol?

5. What part of the cell did the DNA come from?

6. Why can't you see the double-helix structure of DNA?

7. How is a large DNA molecule able to fit within the nucleus of a cell?

B. Summary

In your own words, state what you learned from today's laboratory exercise? What surprised you most? What was the most difficult concept to understand?

Human Anatomy and Physiology

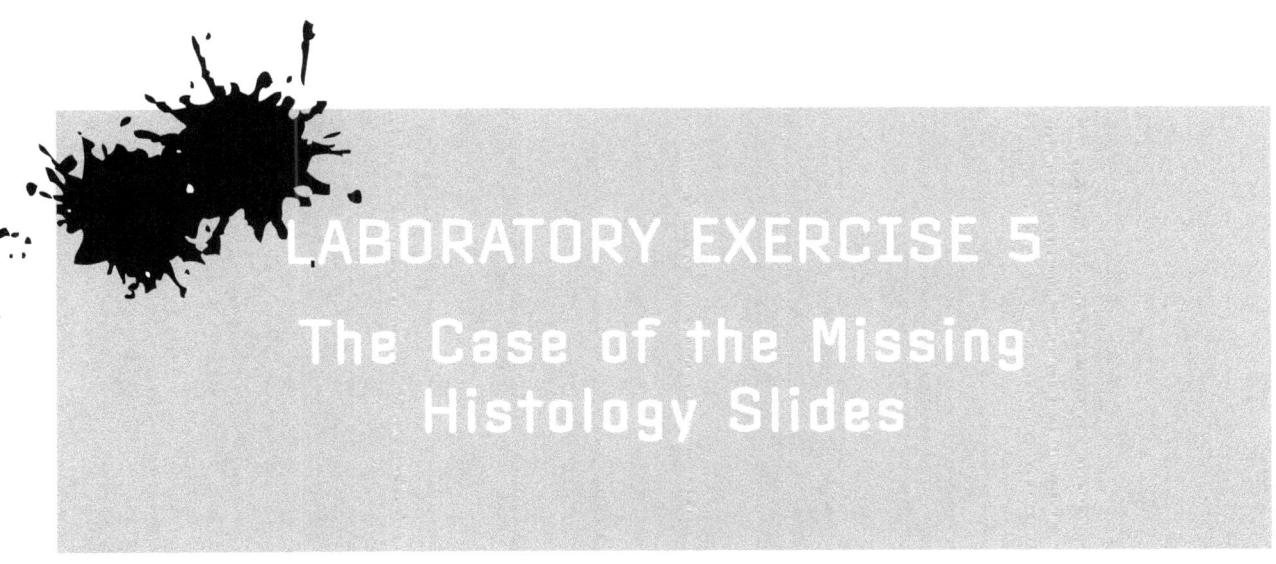

LABORATORY EXERCISE 5
The Case of the Missing Histology Slides

Learning Objective(s)

The student will be able to:

♦ Recognize and identify common tissue types throughout the human body that may be utilized in a crime scene investigation.

MATERIALS: microscope, various histology slides

Your Name: _____   Today's Date: _____

Class: _____   Class Time: (Day) _____ (Time) _____

I. Introduction

Forensic pathologists are trained to analyze soft tissue and organs. In cases in which soft tissue has been degraded by time, temperature, environment or other external forces, the only tissue that may be intact is hard tissue such as bone.

II. The Crime Scene

One cold winter evening at approximately 9:00 pm your instructor, Ms. Sock-It-to-'Em, was in the lab (Room 127) devising the toughest of all lab practicums over tissue types. She had just put the finishing touches on setting up the last microscope and slide sample when she remembered she needed to make some copies for another class she was teaching the next day. Knowing that it was particularly late and no students were likely to be around, she jaunted quickly down the hall to the copier room, leaving the lab door slightly ajar. While she was making copies, the electricity went off and Ms. Sock-It-to-'Em found herself standing in complete and utter darkness. She couldn't even see her hand in front of her face!

After a few minutes, the electricity came back on. Ms. Sock-It-to-'Em finished her copies and swiftly returned to the lab. Upon entering the room, she discovered an unbelievable sight! All of the tissue slides that she had just finished setting up under the microscopes were gone! In fact, a large majority of the microscopes were missing as well. Ms. Sock-It-to-'Em speculated that the thieves must have been looking for the answer key, but she had taken it with her.

As she looked around to access the damage, she noticed numerous broken slides on the ground but five unbroken tissue slides. Apparently, the thieves had dropped them in their haste. There also appeared to be a small trail of blood—perhaps they cut themselves on the glass slides. There was also a spilled box of chicken nuggets and a half-eaten chicken drumstick with exposed gristle at the end of the bone. One old bandaid was crumpled on the floor by the door with what appeared to be a chunk of skin stuck to it. Additionally, there was a typed note taped to the front desk that read, "Down with science! Down with tests!" Ms. Sock-It-to-'Em contacted the campus police immediately and filed an incident report. The police mentioned that they would take a look around and see what they could determine. After a thorough look around not only the lab but the adjacent hallways and outside parking areas, they discovered the tissue slides had been dumped in a hall trash can and the microscopes had been discarded in an outside dumpster.

A. The campus police took samples and set up several exhibits of the evidence. For general viewing with the naked eye:

Your Name: _____   Today's Date: _____

Class: _____   Class Time: (Day) _____ (Time) _____

    Exhibit 1–broken microscopes
    Exhibit 2–typed note
    Exhibit 3–broken microscope slides

B. For microscopic examination:

    Exhibit 4–Sample of blood prepared as a slide

    Exhibit 5–Sample of chicken nugget meat prepared as a slide

    Exhibit 6–Skin sample from bandaid prepared as a slide

    Exhibit 7–Gristle from drumstick prepared as a slide for viewing

    Exhibit 8–Five microscope slides with original tissue still found intact:
       1. Simple squamous epithelial
       2. Cubiodal epithelium
       3. Cardiac muscle
       4. Smooth muscle
       5. Spinal cord (nerve)

III. The Crime Report

You have been asked to assist the case as crime scene technicians. You will need to prepare a crime report. You will be examining eight different exhibits from the crime scene in an effort to find the culprit. For each exhibit you will record drawing(s) of the exhibit as seen either with the naked eye or using the microscope.

A. Crime Report Introduction

Person who reported incident:
Location:
Time:
Describe the nature of the incident:

Your Name: _____   Today's Date: _____

Class: _____   Class Time: (Day) _____ (Time) _____

B. Examination of Evidence

For general viewing with the naked eye (just observe the following)

Exhibit 1–broken microscope(s)
Exhibit 2–typed note
Exhibit 3–broken microscope slides

C. For microscopic examination (prepare sketches of the following evidence)

Exhibit 4–Sample of blood (prepared as a slide)

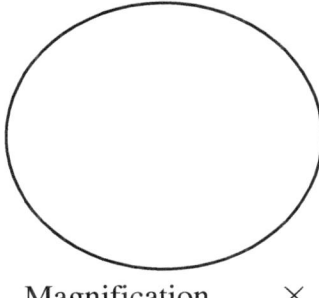

Magnification ____ ×

Exhibit 5–Sample of chicken nugget meat (prepared as a slide)

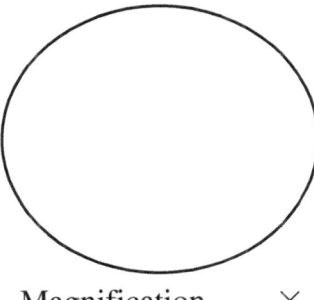

Magnification ____ ×

Exhibit 6–Skin sample from bandaid (prepared as a slide)

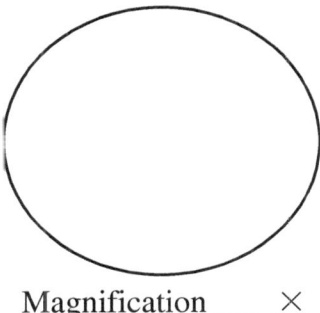

Magnification ____ ×

Laboratory Exercise 5

Your Name: _____ Today's Date: _____

Class: _____ Class Time: (Day) _____ (Time) _____

Exhibit 7–Gristle from drumstick (prepared as a slide)

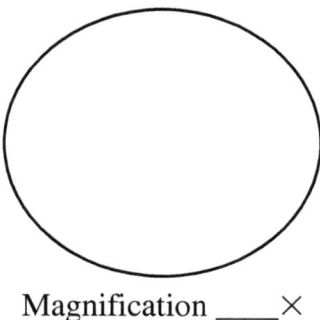

Magnification ____×

Exhibit 8–Five microscope slides with original tissue still found intact

a. Simple squamous epithelial

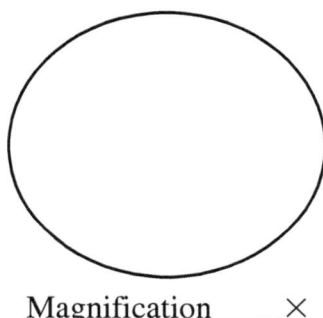

Magnification ____×

b. Cubiodal epithelium

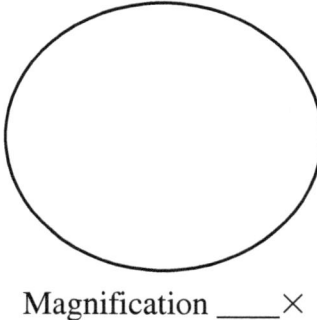

Magnification ____×

Your Name: _____ Today's Date: _____

Class: _____ Class Time: (Day) _____ (Time) _____

    c. Cardiac muscle

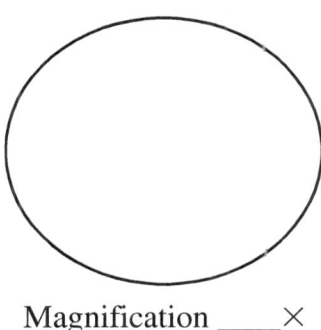

Magnification ____×

    d. Smooth muscle

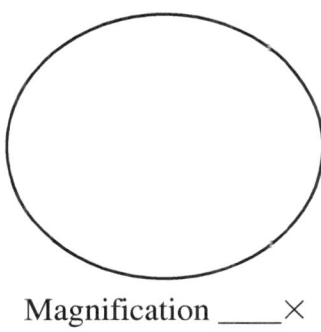

Magnification ____×

    e. Spinal cord (nerve)

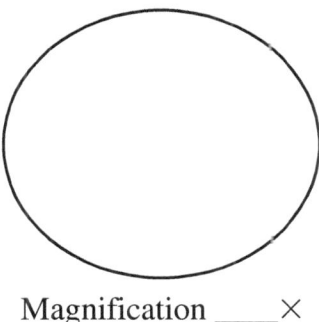

Magnification ____×

Laboratory Exercise 5

Your Name: _____   Today's Date: _____

Class: _____   Class Time: (Day) _____ (Time) _____

D. The Suspects

Four suspects were apprehended by the campus police. All were identified by eyewitnesses as being in the vicinity of the Allied Health building on the evening in question.

Suspect 1: Female age 20 with green and orange highlighted hair wearing dog collar. An iguana tattoo noted on her left forearm. Last seen sitting on a bench at the end of the hall eating food from a box and text messaging on her cell phone. Study materials surrounded her. She had a bandaid on her right hand.

Suspect 2: Male age 19 with dark brown hair wearing gray sweatshirt-type jacket with hood. Appeared to have a textbook zipped up under his jacket. Suspect had ear buds in his ears—perhaps an iPod. Standing near hall trash can. Backpack on ground leaning against trash can. Had tissue or toilet paper wrapped around second or third finger of left hand, which appeared to be bleeding.

Suspect 3: Female age 23 with curly long red hair sitting on floor down the hall. Appeared to be eating finger food. She had typed pages around her on the floor, and she was glancing at them. She had a large tote bag with a small dog who poked its head out to look around. She was feeding the pooch.

Suspect 4: Male age 25 with short spiky blonde hair walking with backpack on his back and backpack in his hand. Wearing shirtsleeve polo-type shirt. Had barbed-wire tattoo around entire left bicep. He was carrying a small plastic party tray of what appeared to be chicken drummettes.

E. Formulating A Hypothesis/Motive

1. Write your thoughts here about a possible motive:

2. What specific type of tissue is the skin (found in the bandaid)?

Human Anatomy and Physiology

Your Name: _____ Today's Date: _____

Class: _____ Class Time: (Day) _____ (Time) _____

   3. What specific type of tissue is the chicken nugget meat?

   4. What specific type of tissue is the gristle on the end of the drumstick?

F. Concluding Questions

   1. What does all the microscopic evidence have in common?

   2. Who did you think is the most likely suspect(s)? Why?

   3. If you could actually question a suspect or two, what would you like to ask them? Write your questions here.

LABORATORY EXERCISE 6
A Closer Look at Skin

Learning Objective(s)

The student will be able to:

- ♦ Understand the role that epithelial cells play in crime scene evidence.
- ♦ Describe the three major regions within the skin and the various layers within the epidermis using a model.

MATERIALS: Slides, cover slips, toothpicks, gentian violet stain, alcohol burner, disinfectant, microscopes, 3D human skin model

Your Name: _____   Today's Date: _____

Class: _____   Class Time: (Day) _____ (Time) _____

I. Introduction

Every time you itch and consequently scratch your skin, dead skin cells, known as epithelials, slough off. In essence, we all shed skin cells as well as hair. It has been said that the average person sheds 1.5 lbs of epithelials each year. In fact, dust is made up of mostly epithelials. As you have previously learned, these tissues are an excellent source of DNA and may be used in crime scene investigations.

In some crimes, a victim may fight or attempt to escape. He or she may scratch the attacker with his or her fingernails, and epithelials may be obtained from under the victim's nails.

As you are probably aware, pets also shed hair and epithelials. Sometimes pet epithelials lead investigators toward a conviction if a match is made between pet hair left behind at a crime scene from a suspect's clothing and the domicile where the pet lives as well.

Predict

1. You have been asked to collect as much epithelial evidence as possible from a crime scene that occurred inside a 3-bedroom, 2-bath home. Name as many places as you can think of that are excellent areas/items to gather epithelials from.

2. What does a typical epithelial look like? Draw or describe what you think it looks like.

NOTE: *There are different types of epithelial tissue. Epithelial tissue from the skin's surface is made up of a hard, tough protein called keratin. Keratin makes the skin's surface appear rough. Using a moisturizer such as lotion temporarily helps the skin's surface to appear softer. Epithelial cells may also be obtained from non-keratinized moist areas such the lining of the cheeks/mouth.*

Your Name: _____ Today's Date: _____

Class: _____ Class Time: (Day) _____ (Time) _____

II. Exploring Epithelials

Working in teams of two, now you will examine an epithelial cell from the inside of your cheek. Using a toothpick, gently scrape the inside of your cheek (scrape gently—do not draw blood) and place the clump of cells on a microscope slide. Add 1 drop of gentian violet stain. Then, place a cover slip on top of the cells. Heat the underside of the slide using a low flame from an alcohol burner to dry/fix the stain on the slide.

A. Place the slide under a microscope. Draw your findings.

Epithelial cell ____× magnification

B. How do your prediction of what you thought an epithelial cell looks like and the actual cell compare? Were the two close? Explain.

NOTE: Clean your microscope slide with disinfectant.

III. Regions of the Skin

Using a skin section model and a key (letter "B" for key), identify the following three regions of skin:

(Check off your progress as you go along)

___Epidermis ___Dermis (Derma) ___Hypodermis (Subcutis)

Your Name: _____   Today's Date: _____

Class: _____   Class Time: (Day) _____ (Time) _____

A. Layers and Specialized Structures Within the Regions of the Skin

(Locate and learn the following structures only)

__ 1. Meissner corpuscle
__ 2. stratum corneum
__ 3. stratum lucidum (clear layer)
__ 4. stratum granulosum
__ 5. stratum spinosum

__ 6. stratum basale
__ 7. papillary layer of dermis

__ 8. reticular layer of dermis
__ 9. sweat gland
__ 10. arrector muscle
__ 11. adipose cell
__ 12. hair (inside hair follicle)
 __ a. cortex of hair
 __ b. medulla of hair
__ 13. sebaceous (oil) gland
__ 14. sweat pore

B. Key to Skin Model.
The letters on this photo correspond to the anatomical parts listed in the previous table

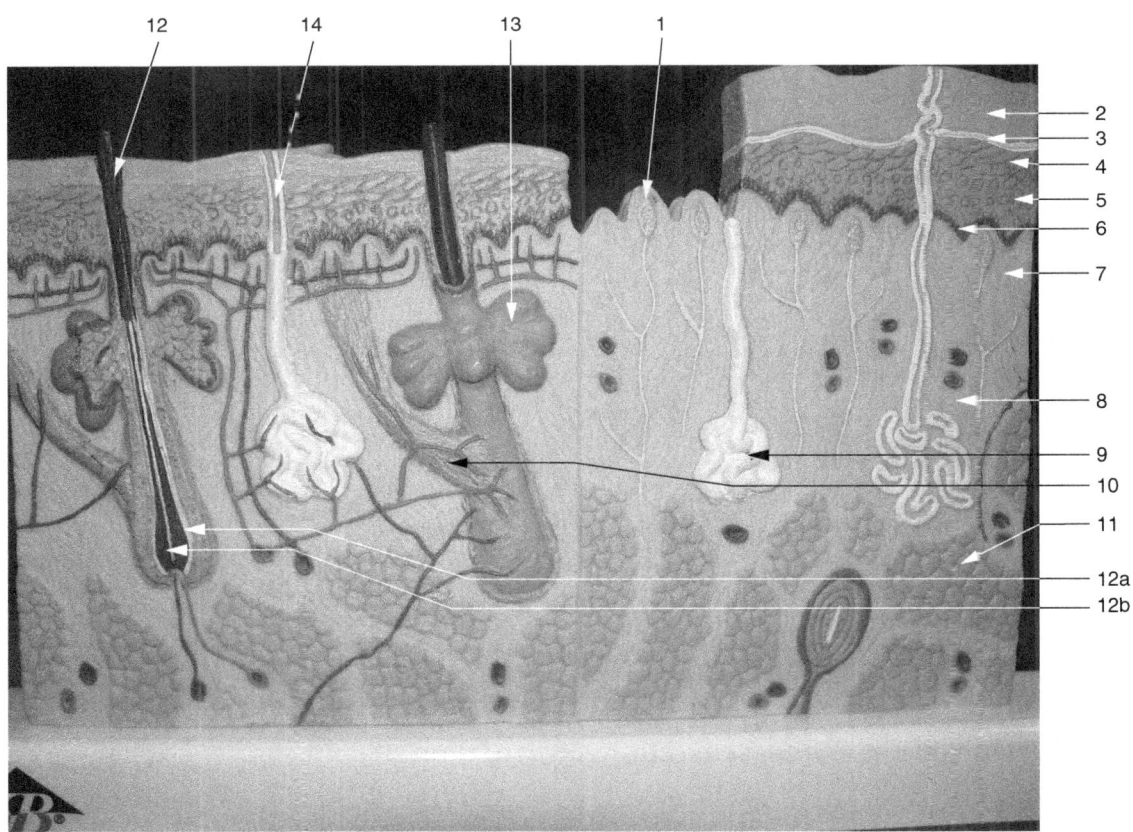

Laboratory Exercise 6

Your Name: _____ Today's Date: _____

Class: _____ Class Time: (Day) _____ (Time) _____

C. Labeling: Practice labeling the skin on the photograph below. CHALLENGE: DO THIS FROM MEMORY!

Here are the structures one more time: *Meissner corpuscle, stratum corneum, stratum lucidum, stratum granulosum, stratum spinosum, stratum basale, papillary layer, reticular layer, sweat gland, adipose cell, hair, hair medulla, hair cortex, arrector muscle, sebaceous gland, sweat pore*

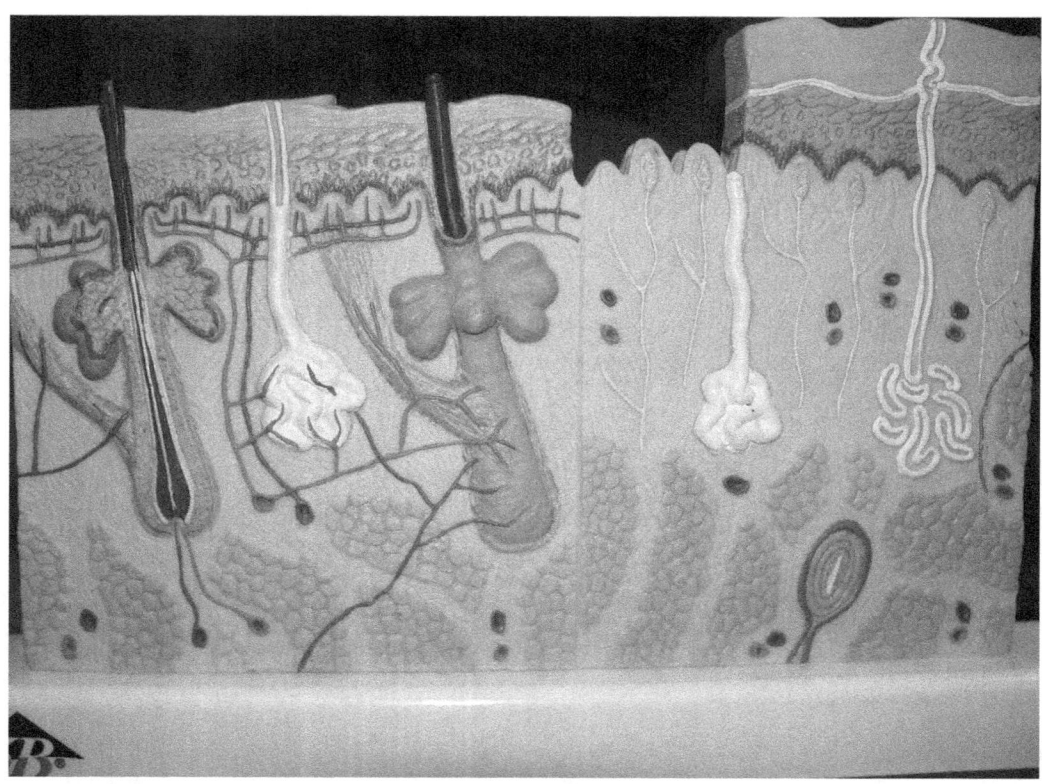

IV. Follow-up Questions

1. Skin cells are shed from what layer of the epidermis?

2. Would you ever expect skin cells to be shed from stratum basale? Explain.

Your Name: _____   Today's Date: _____

Class: _____   Class Time: (Day) _____ (Time) _____

3. You are handling a sheet of paper and receive a shallow paper cut to the stratum corneum, stratum granulosum, and stratum spinosum. You do not bleed. How do you account for this since you were cut across three layers?

4. A 49-year-old woman was physically assaulted. She fought off her attacker by scratching the person with her fingernails. She apparently drew blood. If you were the crime scene investigator examining her nails, what layers or regions of skin would you expect to find? Explain.

5. Explain the difference between keratinized and non-keratinized epithelial tissue. Incorporate the terms "cheek cell" and "skin model" into your discussion.

V. Conclusion

What new information did you learn from today's laboratory exercise?

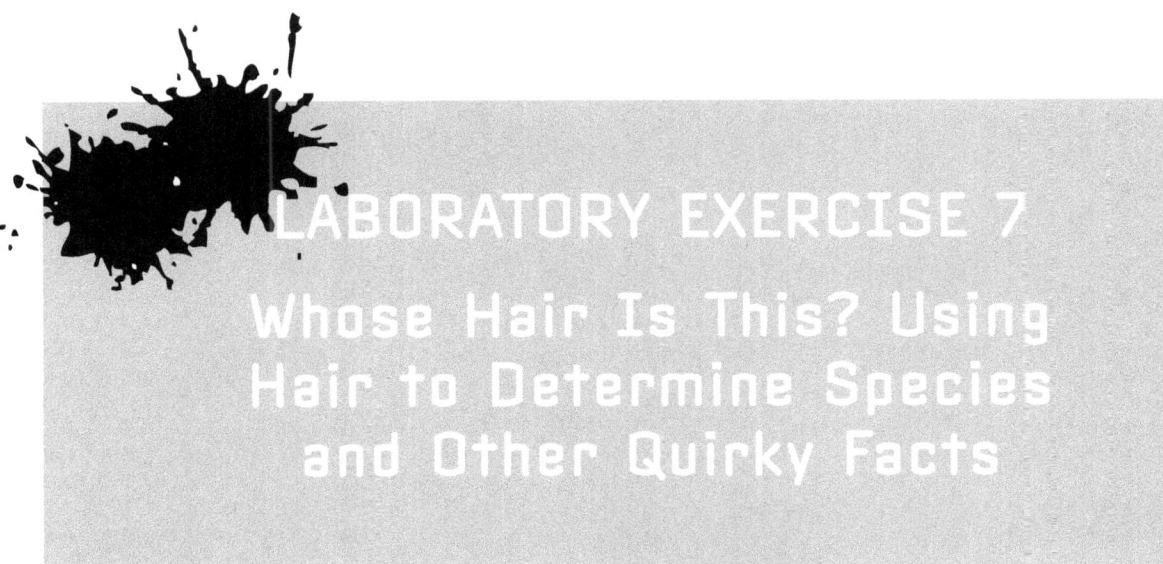

LABORATORY EXERCISE 7

Whose Hair Is This? Using Hair to Determine Species and Other Quirky Facts

Learning Objective(s)

The student will be able to:

- Explain the structure of hair.
- Determine the origin/species of three unknown hair samples using forensic techniques.

MATERIALS: Three unknown hair samples, nail polish, microscope slides and cover slips, mounting medium (for hair to be fixed to slides), clue cards/ background information about hair from different animals (human and otherwise), microscope

Your Name: _____ Today's Date: _____

Class: _____ Class Time: (Day) _____ (Time) _____

I. Background

A. The Use of Hair in Forensic Investigations

Human hair is frequently found at crime scenes, especially violent crimes. It can provide a link between the criminal and his/her act. Hair is considered a "good find" as evidence because it does not decompose as quickly as other human tissues and fluids. But as good as it is to find, it is only considered *class evidence* in a court of law because it is often difficult to link a particular hair sample specifically with a certain person.

Hair samples are not only obtained from the head but also from facial, body, or pubic regions. As you may have observed, hairs from different body regions can have different textures and color.

In order to gain a conviction, crime scene investigators are interested in the matching of color, length, and diameter as well as determining similarities in the medulla (center or core) and cortex (outer layer under the cuticle) of the hair. Additionally, hair has a certain scale pattern. Collectively, scale pattern, medulla, and cortex characteristics are important in hair identification.

Whether the hair has been dyed or bleached can easily be determined. You can actually observe the dyed color on top of the original color under the microscope. Even the approximate time of loss can be determined since it is generally accepted that hair grows about a half inch per month. Additionally, the way the hair was lost can easily be determined. If, for instance, the hair was forcefully removed as in a violent crime, the root of the hair will be large and jagged. A hair that normally falls out due to natural processes has a small smooth root.

Hair has three basic layers: the cuticle, cortex, and medulla. **NOTE:** *The cuticle is composed of scales that cover the hair shaft. The figure below outlines the three distinct layers on a hair shaft.*

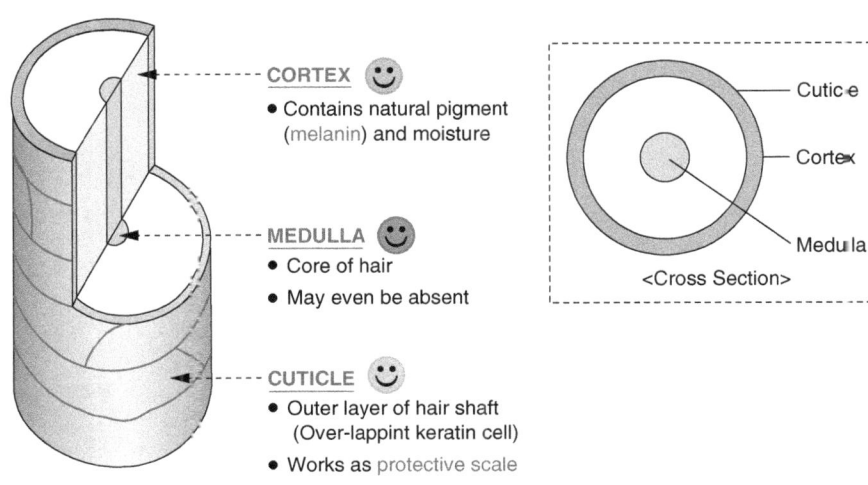

Laboratory Exercise 7

Your Name: _____   Today's Date: _____

Class: _____   Class Time: (Day) _____ (Time) _____

B. Hair Structure

Sometimes, animal hairs are collected from clothing, carpet, or other evidence at the scene of the crime. Perhaps the criminal is the owner of a pet? Perhaps the victim is a pet owner? If like hairs can be matched with both the criminal and the crime scene or victim, the likelihood of conviction is much stronger. Examination of the scale patterns on a hair's shaft play an important role in determining the species.

The scales always point toward the tip of the hair. Basic scale patterns are **coronal**, **spinous**, and **imbricate**.

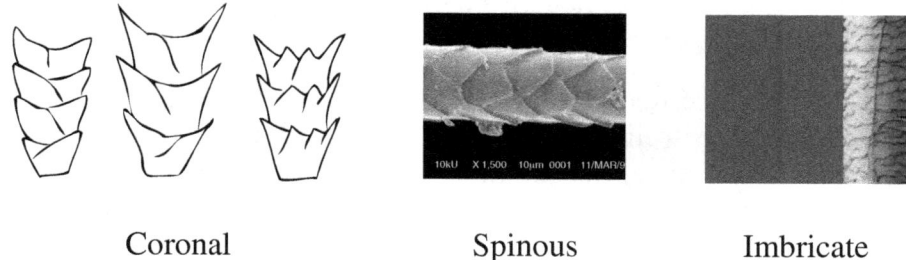

Coronal   Spinous   Imbricate

*See diagram 3 on the "Student Diagram Chart" reference material sheet for more information.

The core or center (medulla) of the hair shaft is not always present, but when present, it shows great variations. This makes the medulla very useful in species identification. The medulla's appearance can be **fragmented**, **intermittent**, or **continuous**.

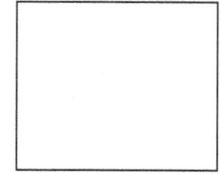

The basic structure of the medulla can also vary. Some of the more common medulla types are uniserial, multiserial, vacuolated, lattice, amorphous (as shown in Diagram 5 on the "Student Diagram Chart").

II. Procedure

A. Observe each hair sample with naked eyes only. Make a hypothesis about each hair type and possible species. What kind of hair do you think it is?

Human Anatomy and Physiology

Your Name: _____   Today's Date: _____

Class: _____   Class Time: (Day) _____ (Time) _____

Sample	Hypothesis About Possible Species of Animal (Human or Otherwise)
A	
B	
C	

B. Preparation of Scale Casts ("Impressions")

1. Coat three microscope slides with a thin layer of nail polish (leave a non-painted margin on the slide to hold onto). Let the polish stand for a few minutes so that it begins to harden. Do not let it dry completely or harden too much (but let it dry enough to leave an impression in it).

2. Label or denote three slides as "A," "B," and "C". Place the appropriate hair sample on each slide (from step 1 above)—<u>allow to dry</u>. **NOTE:** *Leave a piece of the hair extending beyond the edge of the slide.*

3. Remove the hair samples by pulling up on the extended piece of hair.

4. Observe the slides at 10× and 40× (if appropriate/helpful).

5. Sketch the scale pattern (from the cast) for each sample (next page).

Sample A

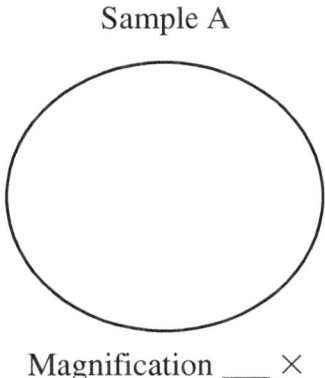

Magnification ___ ×

Laboratory Exercise 7

Your Name: _____   Today's Date: _____

Class: _____   Class Time: (Day) _____ (Time) _____

Sample B

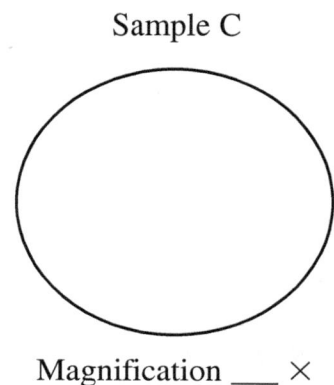

Magnification ___ ×

Sample C

Magnification ___ ×

6. Classify the scale patterns as to their type. Record your findings:

Sample	Findings (patterns, etc.)
A	
B	
C	

NOTE: Discard slides with dried fingernail polish (they cannot be reused).

Your Name: _____ Today's Date: _____

Class: _____ Class Time: (Day) _____ (Time) _____

C. Preparation of Whole Mounts

1. Obtain three clean slides and cover slips. Label A, B, and C.

2. Wet a small surface of the slide with mounting medium. Place the hair sample in this wet area to secure the hair in place. Use two (same) hairs per slide. **NOTE:** *Be careful not to get mounting media on the microscope lenses.*

3. For long hair, wrap hair in figure 8 pattern using three areas of mounting medium (one drop to secure the ends, one drop on middle).

4. Hold cover slip horizontally in hand.

5. Quickly invert cover slip onto slide starting at one edge and lower (like a hinge—not all at once). You are trying to avoid creating air bubbles under cover slip.

6. Observe at 10× and/or 40× (if appropriate/helpful). Examine entire shaft, root (if it exists), variations in color, length, diameter, and medullary characteristics. Record your sketches and findings for samples A, B, and C.

Sample A–Noted Characteristics: _____

_____

_____

Your Name: _____ Today's Date: _____

Class: _____ Class Time: (Day) _____ (Time) _____

Sketch–Sample A

Sample B–Noted Characteristics: _____

_____

_____

Sketch–Sample B

Sample C–Noted Characteristics: _____

_____

_____

Your Name: _____   Today's Date: _____

Class: _____   Class Time: (Day) _____ (Time) _____

Sketch–Sample C

D. Data Summary

Based on your microscopic observations and findings, what is your group's conclusion now?

Sample	Species Identified	Supporting Reasons
A		
B		
C		

E. Exploring Further (this section optional)

Study the root sketches (Diagram 1) on the "Student Diagram Chart." Compare the sketch of the human hair that was forcibly removed from the scalp (note the portion of the follicle that was taken with it) versus the hair that simply fell out (a club-shaped root).

1. Pluck a hair from your head and prepare a whole mount slide.

2. Study the root of the hair.

3. Sketch what you see here:

Sketch of Forcibly Removed Hair Root ___×

Your Name: _____ Today's Date: _____

Class: _____ Class Time: (Day) _____ (Time) _____

F. Conclusion

1. Why might it be helpful for criminologists to create a cast when studying human hair (say as opposed to animal hair)?

2. What implications might exist from a crime scene where large numbers of forcibly removed hair strands are found?

3. Explain the structure of hair in your own words.

4. Using Diagram 5 on the Student Diagram Chart as a reference, explain the medullary differences between rabbit, deer, and dog/fox hair.

5. Outline the use of hair as *class evidence* and how a CSI might justify hair samples as "solid" evidence in a court of law.

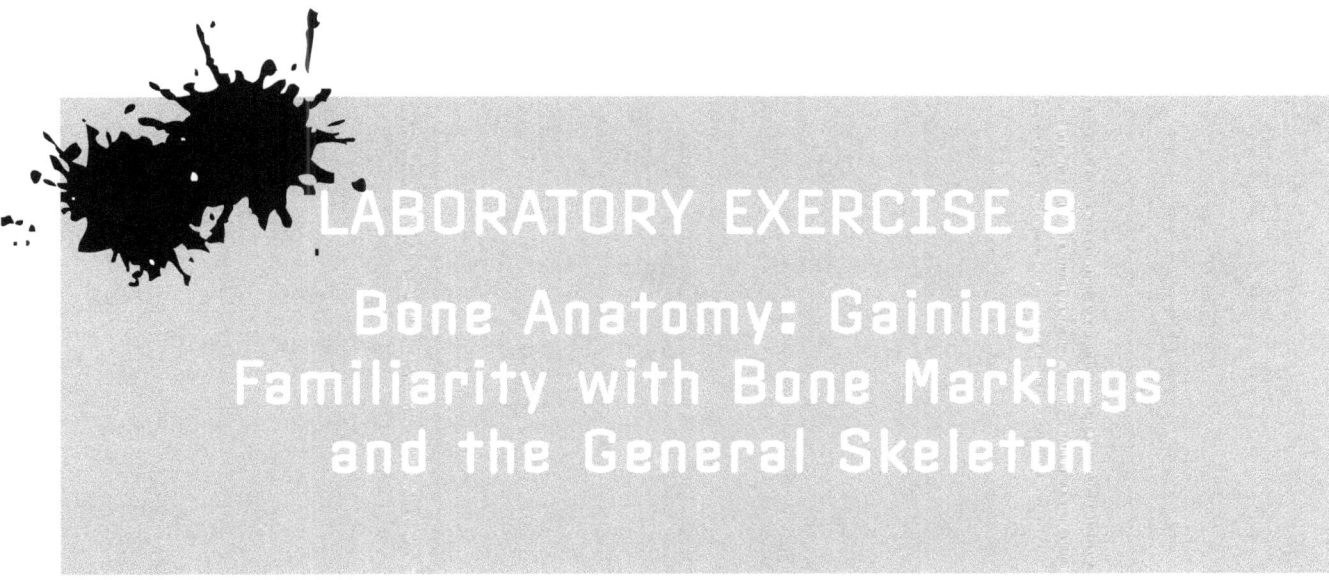

LABORATORY EXERCISE 8

Bone Anatomy: Gaining Familiarity with Bone Markings and the General Skeleton

Learning Objective(s)

The student will be able to:

- Study bone shapes and bone markings. Be able to identify on skeleton models.
- Identify the majority of the 206 bones of the human body.

Your Name: _____   Today's Date: _____

Class: _____   Class Time: (Day) _____ (Time) _____

I. Introduction

Bone Appearance

1. Select a bone to observe closely—any bone such as a leg bone, hip, or shoulder blade will do. Beyond the size or shape of the bone, what do you notice? For example, is it smooth or rough? Write your response here.

   As you may have observed, bones are not smooth. They have projections, depressions, holes, bumps, etc. These are known as **bone markings**. Bone markings give tendons and ligaments a place to "grab onto." They also provide important passageways for nerves and blood vessels.

2. Bones are classified into two main groups. Fill in answers below:

   a. the _____ skeleton consisting of the skull, vertebrae, and ribs

   b. the _____ skeleton consisting of the limbs, shoulder, and hip areas

II. Bone Shapes.
Bones are further classified by their basic shapes. Complete the table below:

Bone Shape and Description	Bone Sketch (Draw one example for each shape)
a) Long bones—longer than they are wide such as the bones of the upper arm or upper leg	a) Name of this bone(s):
b) Short bones—cube-shaped bones of the wrist and ankle	b) Name of this bone(s):

Laboratory Exercise 8

Your Name: _____   Today's Date: _____

Class: _____   Class Time: (Day) _____ (Time) _____

Bone Shape and Description	Bone Sketch (Draw one example for each shape)
c) Flat bones—thin, flattened, and a bit curved as in the breast bone and most skull bones	c) Name of this bone(s):
d) Irregular bones—bones with complicated shapes (e.g., vertebrae and hip bones)	d) Name of this bone(s):

III. Bone Markings. As you have seen thus far, bones are rarely smooth. They have projections, depressions, and openings called **bone markings**. And each bone marking has a specific anatomical name (i.e., facet).

Instructions: Using the table below, **familiarize yourself with each bone marking, its description, and location on the skeleton**. Check off your progress as you go along:

Bone Marking	Description and Location
__ 1. facet	Typically seen on a vertebrae—a smooth, flat, articular (adjoining; abutting) surface. Sketch example:
__ 2. foramen	Typically seen on the jaw—a round or oval opening through a bone. Sketch example:

Your Name: _____   Today's Date: _____

Class: _____   Class Time: (Day) _____ (Time) _____

Bone Marking	Description and Location
__ 3. trochanter	Only seen on the upper leg bone—a blunt, irregularly shaped process. Sketch example:
__ 4. process	Any bony prominence. Sketch example: Where is this located?
__ 5. sinus	Air-filled cavity lined with a mucous membrane within a bone (as seen in the skull). Sketch example:
__ 6. crest	Typically seen on the pelvis—a narrow ridge of bone that is usually prominent. Sketch example:
__ 7. head	An example is the upper leg bone—a bony expansion carried on a narrow neck. Sketch example:
__ 8. ramus	Seen on the lower jaw—an armlike bar of bone. Sketch example:

Your Name: _____   Today's Date: _____

Class: _____   Class Time: (Day) _____ (Time) _____

Bone Marking	Description and Location
__ 9. tubercle	Seen on upper leg bone—small rounded projection or process. Sketch example:
__ 10. tuberosity	Seen on the lower arm bone in line with the thumb—large rounded projection; may be rough. Sketch example:
__ 11. fossa	Found where front teeth insert—a shallow basin-like depression in a bone. Sketch example:
__ 12. fissure	Seen in the eye orbits—a narrow slit-like opening. Sketch example:
__ 13. meatus	As seen in the ear canal—a canal-like passageway. Sketch example:
__ 14. epicondyle	Seen on the upper leg bone—raised area on or above a condyle. Sketch example:

Your Name: _____ Today's Date: _____

Class: _____ Class Time: (Day) _____ (Time) _____

Bone Marking	Description and Location
__ 15. line	Seen on the pelvis—a narrow ridge of bone less prominent than a crest. Sketch example:
__ 16. groove	Seen on the costal area of the ribs—a furrow. Sketch example:
__ 17. spine	As seen on a vertebrae—sharp, slender pointed projection. Sketch example:
__ 18. condyle	Rounded articular projection (typically seen on the distal end of the upper leg bone). Sketch example:

IV. Bone Anatomy. Identify the following bones on a skeletal or skull model; check off your progress on the table as you progress. **NOTE:** *Reference the human atlas at the back of the manual.*

Your Name: _____ Today's Date: _____

Class: _____ Class Time: (Day) _____ (Time) _____

Skeleton—Generalized	Skull (including facial bones)
__skull __mandible __sternum __manubrium __xyphoid process __vertebrae __ribs (1–12 including true ribs, false ribs and floating ribs) __costal cartilage __clavicle __scapula __humerus __radius __ulna __carpals __metacarpals __phalanges (digits) __pelvis __femur __patella __tibia __fibula __metatarsals __tarsals	__frontal __parietal __temporal __occipital __occipital condyle __ethmoid __sphenoid (including greater wings, lesser wings and sella tersica) __lacrimal __zygomatic __maxilla __mandible __pterygoid processes __mastoid processes __styloid processes __palatine __vomer __foramen magnum __external auditory canal
Vertebral Column with Pelvic Girdle	Pectoral Girdle
__cervical vertebrae (C1–C7; indicate spinous process) __thoracic vertebrae (T1–T12) __lumbar vertebrae (L1–L5) __sacrum __coccyx __pubis __pubic symphysis __ilium __ishium (including obturator foramen) __acetabulum	__scapula __acromion __body of scapula __glenoid fossa __coracoid process __clavicle

Instructions: Label the bones from the table (above) on the photos. It is ok to place the bone name on the photo that provides the best view of the bone.

Human Anatomy and Physiology

Your Name: _____   Today's Date: _____

Class: _____   Class Time: (Day) _____ (Time) _____

Skeleton - Generalized

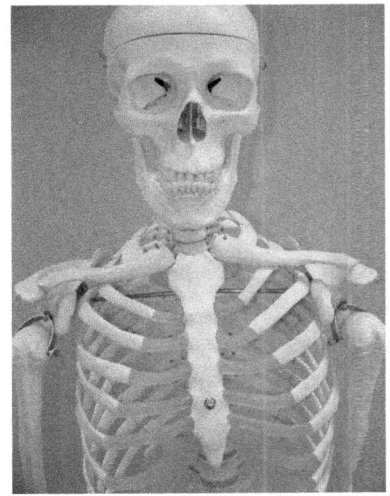
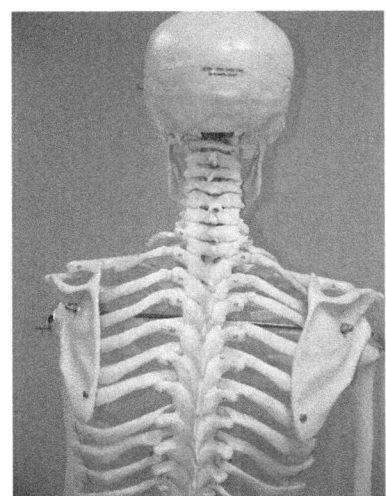
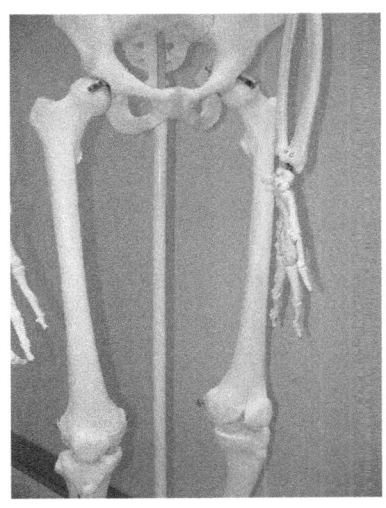
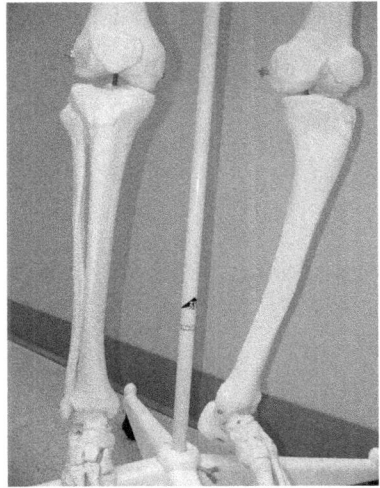

Arm

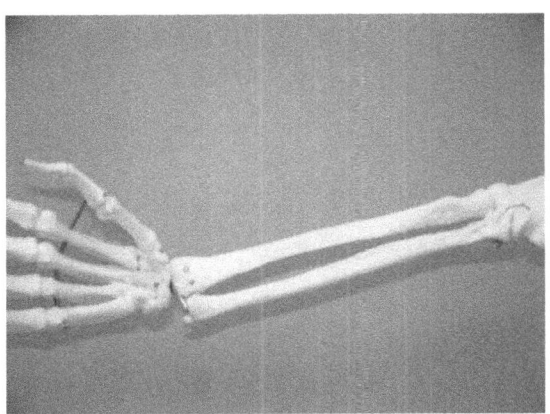

Laboratory Exercise 8

Your Name: _____   Today's Date: _____

Class: _____   Class Time: (Day) _____ (Time) _____

Pectoral Girdle (Posterior Aspect)   Pectoral Girdle - Anterior Aspect

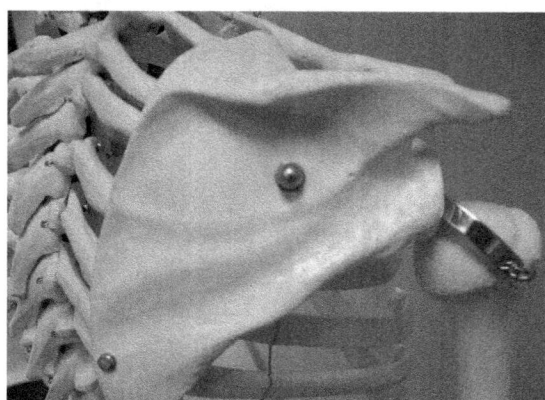

V. Analysis

A. Follow-up Questions

1. What is the largest foramen in the skull?

2. How can you easily distinguish between the radius and the ulna?

3. How can you easily distinguish between the tibia and the fibula?

4. What does articulate mean? (as in the two bones "articulate")

5. How many of each type of vertebrae are in the spine? Cervical? Thoracic? Lumbar?

Your Name: _____ Today's Date: _____

Class: _____ Class Time: (Day) _____ (Time) _____

6. What are some differences that you observe in the spines of cervical, thoracic, and lumbar vertebrae?

7. What is the main difference between true and false ribs?

B. Conclusion

1. State in your own words why bone markings are significant.

2. State what you have learned as a result of this laboratory exercise.

Skeletal Key

Skeleton - Generalized

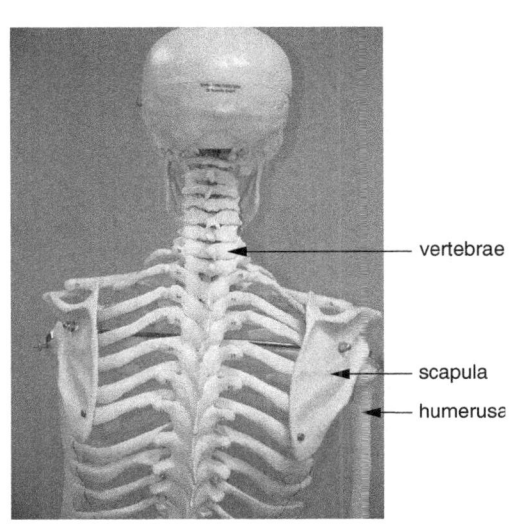

Laboratory Exercise 8

Your Name: _____ Today's Date: _____

Class: _____ Class Time: (Day) _____ (Time) _____

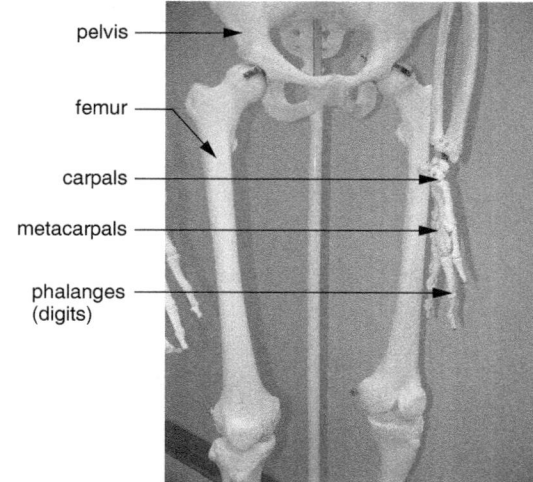
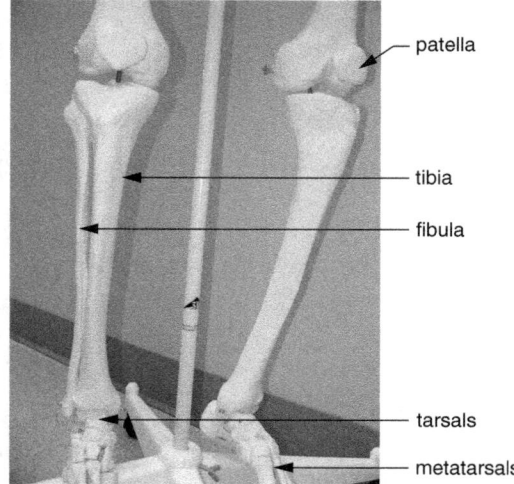

Arm

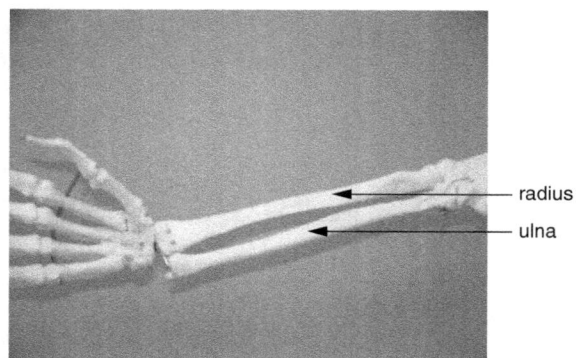

Pectoral Girdle (Posterior Aspect)

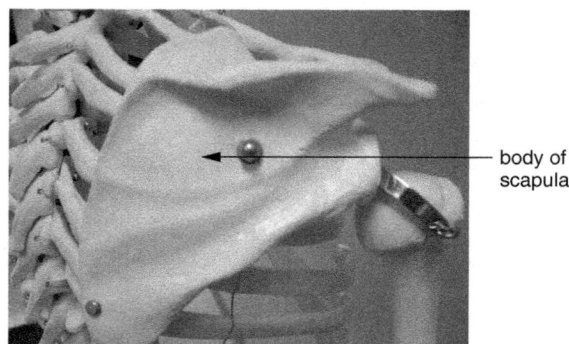

Pectoral Girdle - Anterior Aspect

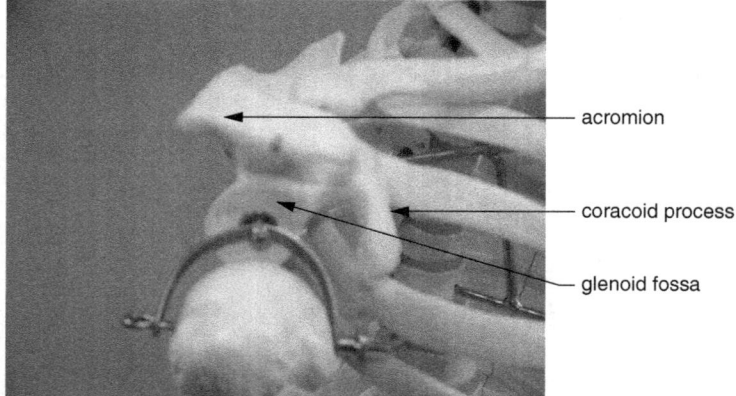

Human Anatomy and Physiology

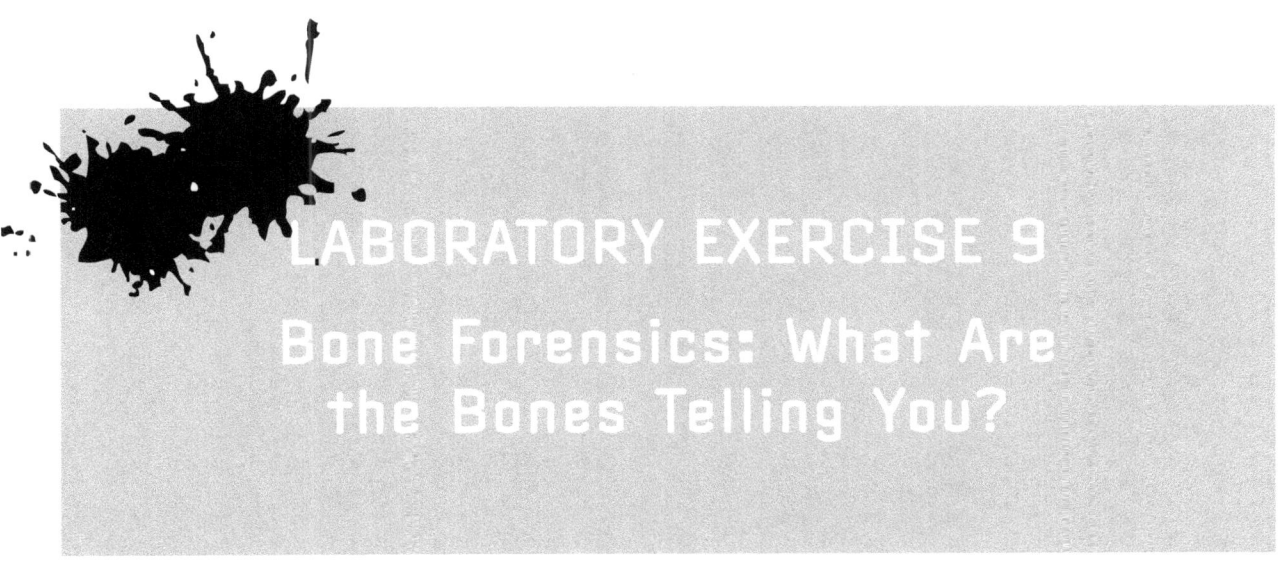

LABORATORY EXERCISE 9

Bone Forensics: What Are the Bones Telling You?

Learning Objective(s)

The student will be able to:

- Explore disarticulated skeletal evidence to determine how many bones you can identify (by anatomical name).
- Determine as much forensic information from your disarticulated skeleton as possible.

MATERIALS: disarticulated skeleton, ruler, calculator

Your Name: _____   Today's Date: _____

Class: _____   Class Time: (Day) _____ (Time) _____

I. Introduction: Skeletal "Anthropology" (Forensics)

Background

Scientists must first determine if the skeletal remains are animal or human. If human, they must then try to determine growth rate, gender, and height of person.

A. Age of a Skeleton. Teeth that have or have not grown can also reveal the age of the skeleton, as young children will have not lost their milk teeth, and at the age of 18, wisdom teeth first appear. During the teenage years, bones become thicker and larger and fuse together in a process known as "ossification." Ossification occurs in 800 points of the body and is the best guide to revealing the age of a child's skeleton. An example of ossification occurs in the arms, where at age 6, the two bone plates form at either end of the outer forearm (radius).

At age 17 in males and 20 in females, the lower bone plate and the radius fuse together; soon after, the upper bone plate and radius fuse together. The bone in the body that finishes growing last is the collarbone, which ceases growth at 28 years. In the bones of the elderly, degeneration begins to occur. Anthropologists will look for tiny spikes that start to appear on the edges of the vertebrae, the wearing of teeth due to age and joints that show signs of arthritis. All of the bones in the body will deteriorate with age.

B. Gender. When determining male and female in a skeleton, anthropologists look at the skull and hip bones as they contain clues to the sex of the skeleton.

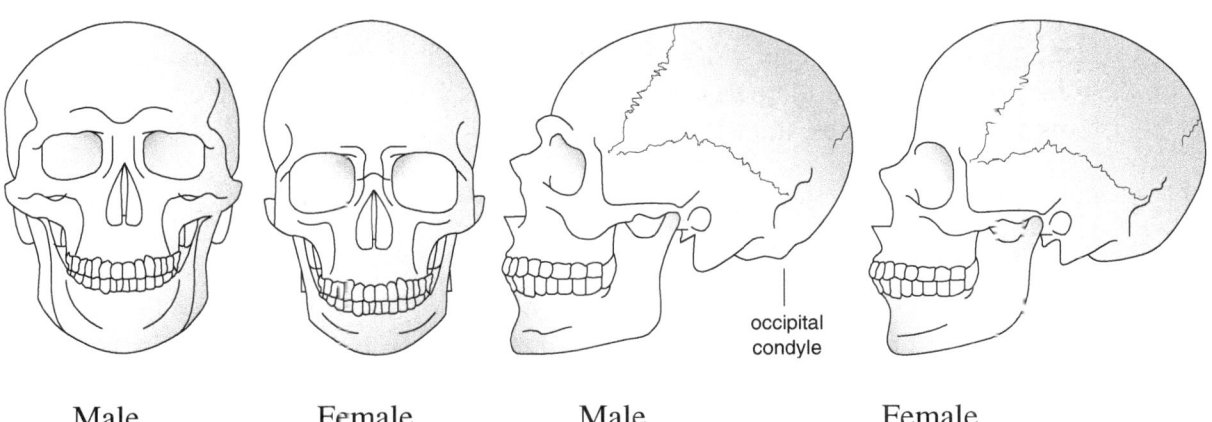

Male   Female   Male   Female

Laboratory Exercise 9

Your Name: _____   Today's Date: _____

Class: _____   Class Time: (Day) _____ (Time) _____

1. **Sexual differences in cranial morphology**

    a. General architecture: In males, the overall construction of the skull is heavier and more rugged looking than that of the female skull.

    b. Eye openings: The orbits in the skull for the eyes are a bit squared in males, while in females they are more rounded.

    c. Brow ridges: The supraorbital ridge of males is heavier and more pronounced than it is in females; in females the brow is smooth and flat.

    d. Cheekbones: The cheekbones of males are heavier and more laterally arched; in females, the cheekbones are lighter, more compressed, and they tend to lack the lateral arching.

    e. Occipital condyle: In males, the occipital bump at the rear base of the skull tends to be much more pronounced than it is in females, where it can be almost nonexistent.

    f. Chin shape: The shape of a male's chin approximates the letter U, a female's the letter V.

    g. Jaw line: The angle where the horizontal portion of the jaw curves upward into the ramus, or vertical part of the jaw, is much more angular in males than it is in females.

2. **Sexual differences in pelvic structure**

    a. General architecture: The width of the pelvic girdle is broader in females than it is in males. In females, the pelvic girdle surrounds a birth canal large enough for the fetus to pass. In males, the pelvic opening is less round and open.

    b. Pelvic opening: The opening of the pelvis, called the pelvic inlet, is rounder and larger in females, while in males it tends to be narrow and more constricted.

    c. Pubic arch: The joining of the bones at the bottom of the pelvis forms a broad angle in females, usually greater than 90°, while in males it is narrow, usually less than 90°.

    d. Pubic bone: In males, the pubic bone is roughly triangular in shape; in females, it is rectangular.

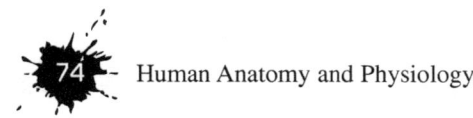

Your Name: _____ Today's Date: _____

Class: _____ Class Time: (Day) _____ (Time) _____

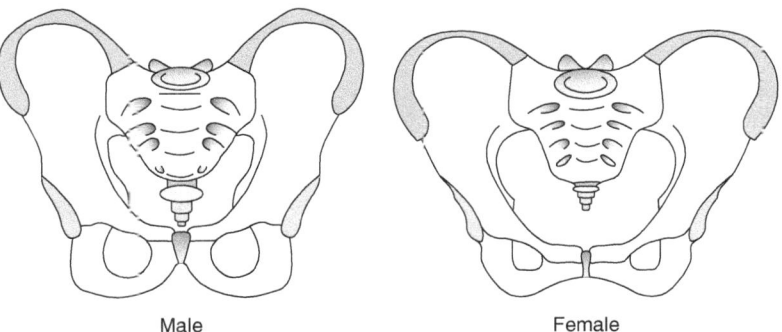

Male   Female

C. Height. Determining the height of a skeleton involves reassembling the skeleton and measuring the length of significant bones. Adding 10–11cm or 4 inches onto the bone length accounts for the missing tissue and muscle. If parts of the skeleton are missing, certain individual bones are used as a height guide. The longer the bone is, the better and more accurate the estimate will be, so the femur is measured first. The human height measures roughly 3 times the length of the femur, although it also depends on the race and sex of the skeleton. So, if the femur bone is 2 ft (2ft × 3) or roughly 6 ft tall; don't forget to add approximately 4 inches for to account for missing tissue and muscle.

II. Procedure

A. Divide into small groups of 3–4 people. Form a forensics team by assuming the following roles within the group:

1. Chief of Forensics. Obtains evidence and returns evidence (clears up); steers the investigation.

2. Lead Investigator. Helps group/organize bones; coordinates the "putting together "of the bones into a recognizable body (lay in out like a puzzle you are putting together).

3. Forensic Reporter. Ensures accurate findings and calculations (someone good with math and accuracy); helps other team members understand concepts.

4. Forensic Artist/Photographer. Makes accurate sketch of assembled bones; assists other team members with their sketches.

B. Chief of Forensics obtains a "bone box" of skeletal evidence (aka: a disarticulated skeleton). Team begins work. Assume that the disarticulated skeleton is the remains/evidence from a crime scene. Use observation and exploration to determine as much as possible about the "evidence" (skeletal remains).

Your Name: _____   Today's Date: _____

Class: _____   Class Time: (Day) _____ (Time) _____

III. Data and Analysis

A. List the bones that you are able to **positively** identify here:

B. How many bones (total) did your group identify? ____

C. Sketch of bones identified (assembled as a body):

IV. Evidence and Conclusion

What can you determine from the bones?

A. Age estimate? Explain.

B. Gender? Why do you think is the gender? Explain.

C. What is the estimated height of the individual? (complete table)

Inferred Height from Femur

_____	Inferred Height from Femur	Actual Height Estimation
Subject #1		

Show Calculation Here (Estimating Height from Femur)

Length of Femur = _____ × _____ = _____

Your Name: _____ Today's Date: _____

Class: _____ Class Time: (Day) _____ (Time) _____

V. Follow-Up Questions

1. How accurate were you in inferring height from femur length? Explain.

   _____

   _____

   _____

2. How would you distinguish a male from a female skull?

   _____

   _____

   _____

   _____

3. How would you distinguish a male from a female pelvis?

   _____

   _____

   _____

   _____

VI. Summary and Conclusion

What did you learn from today's forensic lab? What was most beneficial/interesting/etc.? (Maybe something you did not already know?)

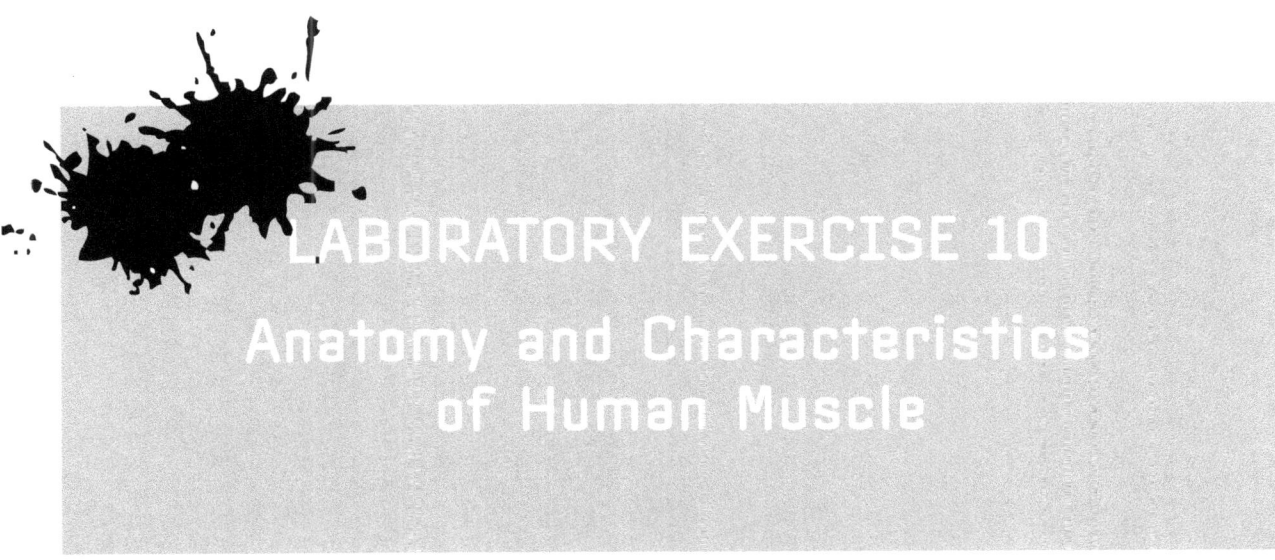

LABORATORY EXERCISE 10
Anatomy and Characteristics of Human Muscle

Learning Objective(s)

The student will be able to:

- Identify the major muscles of the human body on a human model. Think "pseudo-autopsy" today.
- Troubleshoot 3 crime scenes involving stab wound scenarios to determine what type of muscle tissue was involved.

MATERIALS: human anatomical model, slides of various muscle tissue, microscope

Your Name: _____ Today's Date: _____

Class: _____ Class Time: (Day) _____ (Time) _____

I. Gross Anatomy

There are over 700 muscles in the human body. However, you will not learn that many in first-year anatomy.

How would you rate your knowledge of muscle anatomy and muscle tissue types prior to this lab? (1 is least amount of knowledge and 10 is best) Indicate your rating below.

_____
1   5   10

Below is a concise list of the muscles that you are required to learn and identify on a human anatomical model. **NOTE:** Some muscles are not identifiable or easily seen on the models (i.e., platysma, tranversus abdominus, gluteus medius) and are indicated with an asterisk (*). In these cases, try to approximate and describe where the muscle would normally appear.

A. In small teams, find the muscles listed below **on an anatomical model**. Check off your progress as you go along.

Head and Neck
__Temporalis
__Masseter
__Epicranius, frontal belly
__Epicranius, galea aponeurotic
__Epicranius, occipital belly
__Buccinator
__Sternohyoid*
__Zygomaticus, major and minor
__Sternocleidomastoid
__Trapezius
__Platysma*
__Obicularis oculi
__Obicularis oris

Gluteal
__Gluteus maximus
__Gluteus medius*

Leg
__Sartorius
__Rectus femoris
__Tibialis anterior
__Gastrocnemius
__Iliotibial tract
__Achilles tendon

Shoulder, Thorax, and Abdomen
__Deltoid
__Infraspinatus
__Teres major
__Rhomboid major*
__Lattisimus dorsi
__Pectoralis major
__Pectoralis minor*
__Serratus anterior
__Intercostals
__Rectus abdominus
__External oblique
__Internal oblique*
__Transversus abdominus*

Arm
__biceps brachii
__triceps brachii
__brachioradialis
__brachialis

Your Name: _____   Today's Date: _____

Class: _____   Class Time: (Day) _____ (Time) _____

B. Label your practice photographs (below) with the muscles (listed on the previous page ONLY) as reinforcement practice. Label each muscle on the most appropriate photograph. **NOTE:** A key is located near the end of the laboratory exercise for reference.

Head (Anterior aspect)

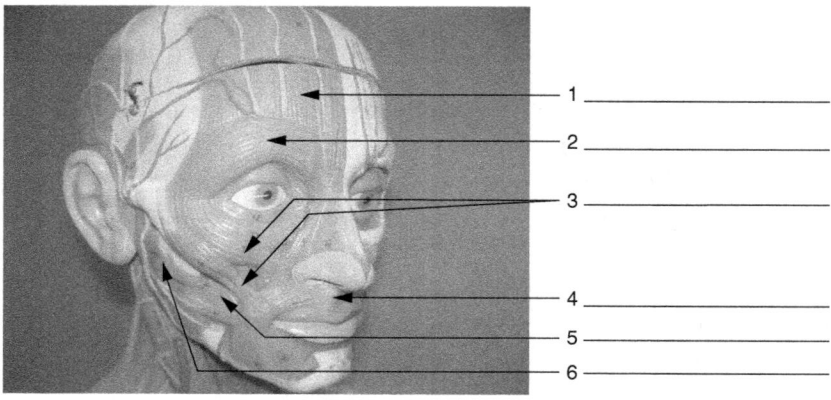

1 _____
2 _____
3 _____
4 _____
5 _____
6 _____

Head (Superior aspect)

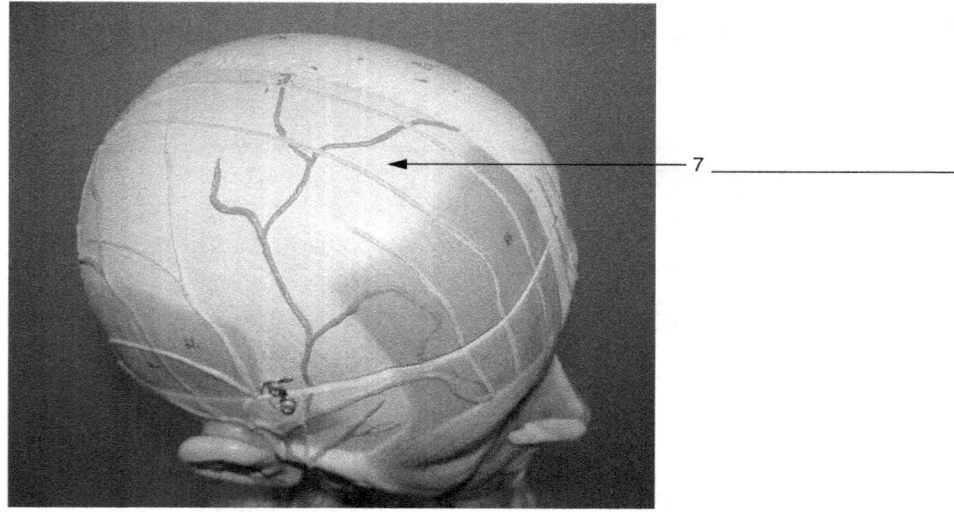

7 _____

Human Anatomy and Physiology

Your Name: _____ Today's Date: _____

Class: _____ Class Time: (Day) _____ (Time) _____

Neck (Anterior lateral aspect)

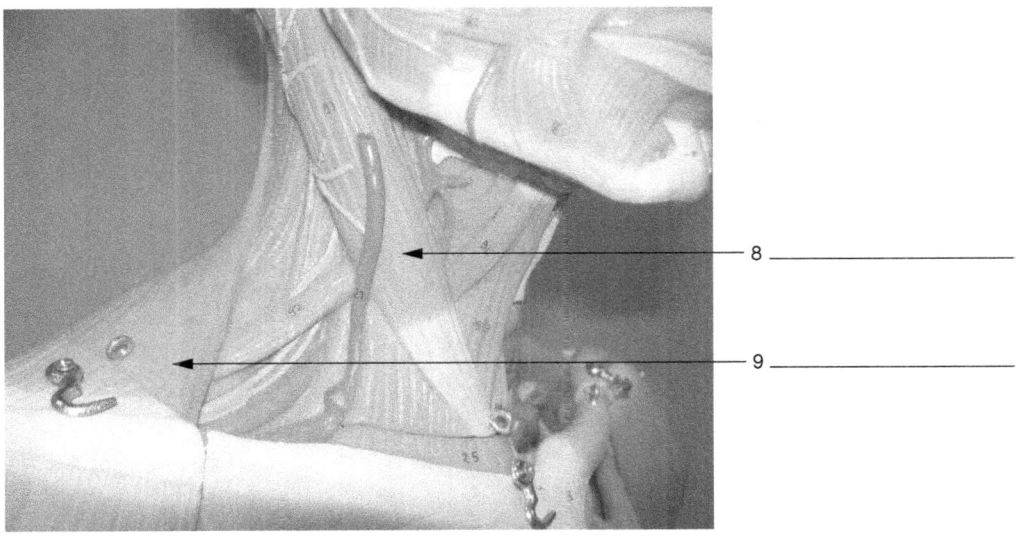

8 _____

9 _____

Thoracic (Anterior aspect with shoulder and arm)

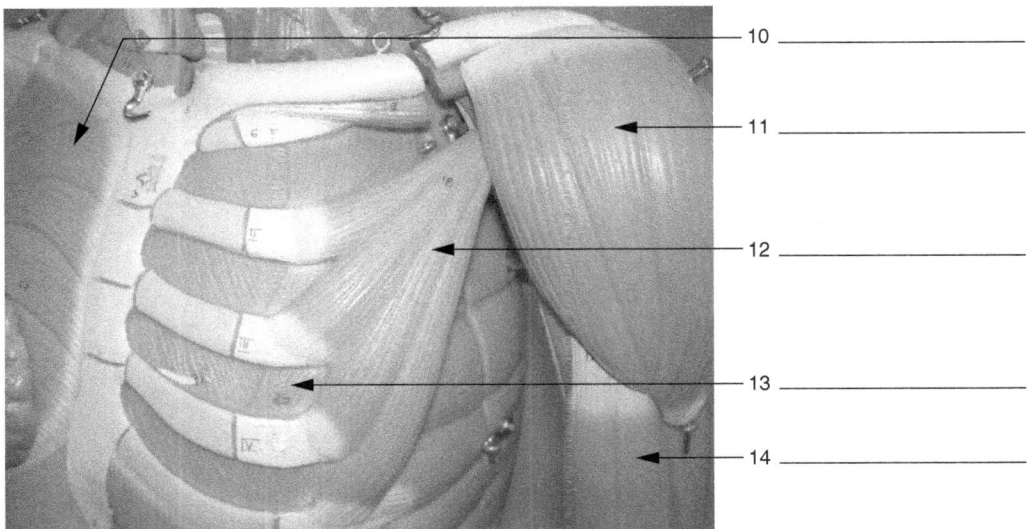

10 _____

11 _____

12 _____

13 _____

14 _____

Laboratory Exercise 10

Your Name: _____ Today's Date: _____

Class: _____ Class Time: (Day) _____ (Time) _____

Thoracic (Posterior aspect with shoulder and arm)

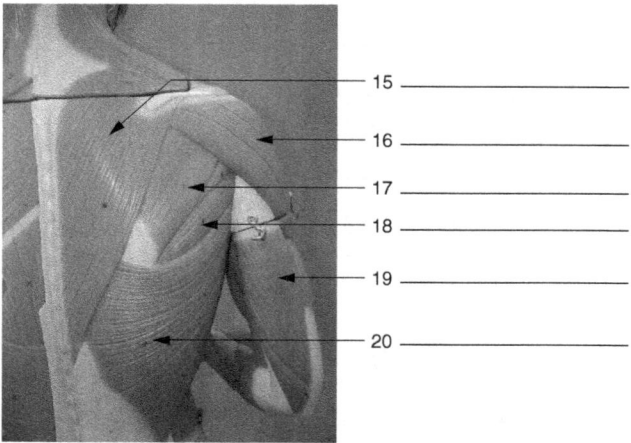

15 _____
16 _____
17 _____
18 _____
19 _____
20 _____

Thoracic (Anterior aspect featuring abdominal area)

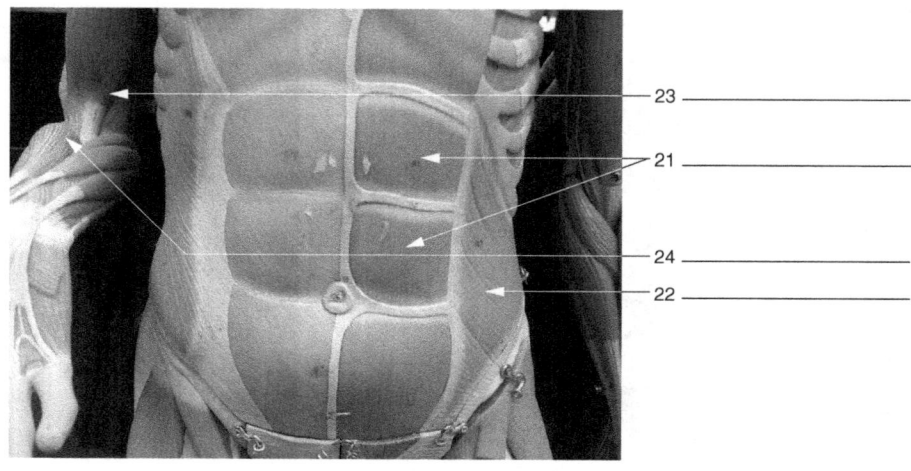

23 _____
21 _____
24 _____
22 _____

Shoulder and Upper Arm (Lateral aspect)

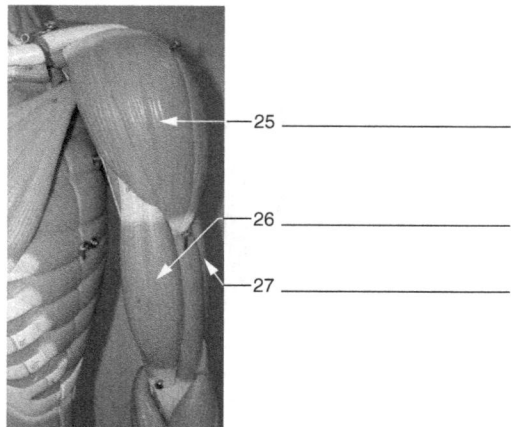

25 _____
26 _____
27 _____

84 Human Anatomy and Physiology

Your Name: _____ Today's Date: _____

Class: _____ Class Time: (Day) _____ (Time) _____

Leg (Anterior aspect)

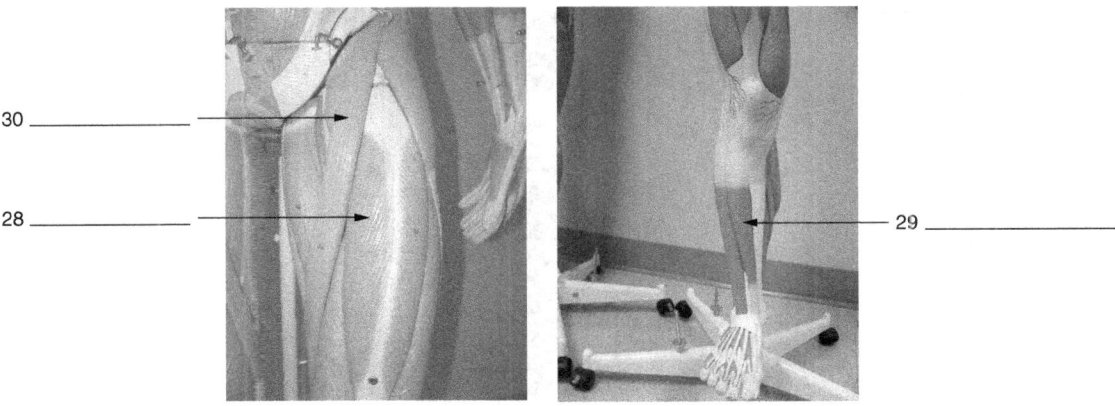

30 _____
28 _____
29 _____

Leg and gluteal area (Posterior aspect)

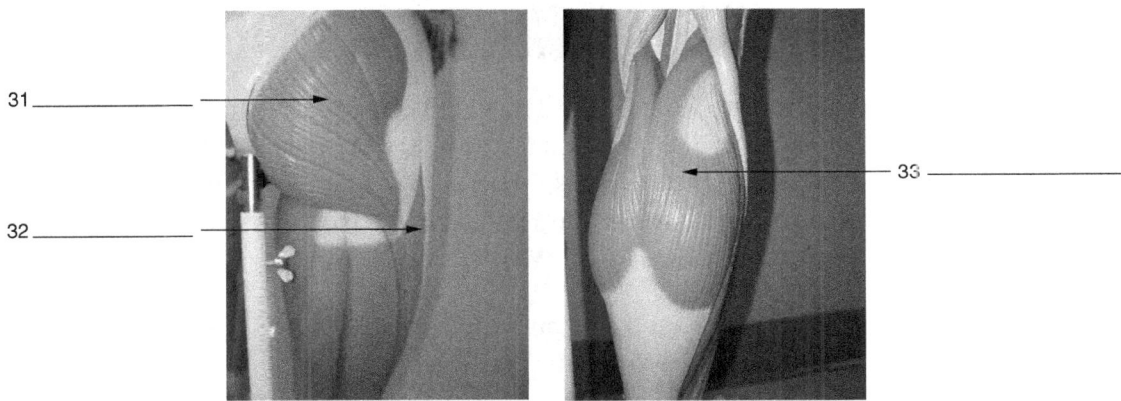

31 _____
32 _____
33 _____

II. CSI – Stab Wound Scenes and Muscle Tissue Types

Instructions: Study each crime scene report to determine the muscle tissue type that is involved. Examine the forensic evidence (that was collected from the crime scene) under a microscope.

Before you begin, create a CSI badge so that you can enter the crime scene areas. Design a team badge which includes your team or unit's name, logo, and badge number 4100325.

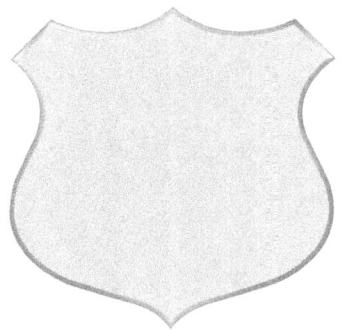

Laboratory Exercise 10

Your Name: _____   Today's Date: _____

Class: _____   Class Time: (Day) _____ (Time) _____

A. Crime Scene #1—Male victim. Case no. 11466860

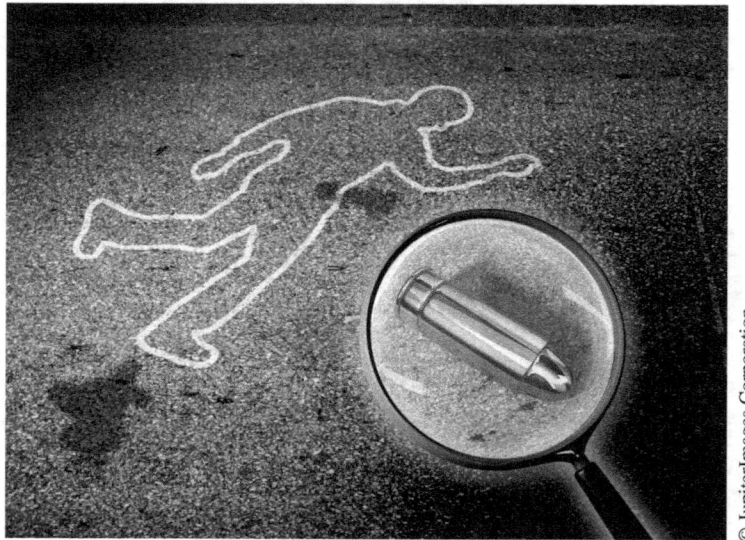

Description: The victim was found in a left lateral position ventral plane facing slightly upward. Noteworthy is a large red stain on the ground next to the body emerging from what appears to be the abdominal area. A 10" knife was also found near the body on the ground. Tissue was retrieved from the blade and prepared for microscopic examination in Exhibit A (slide). The victim apparently died of his wound(s).

1. Probable hypothesis about tissue on knife blade? (What is your hunch regarding the tissue? Likely type of muscle tissue?)

2. Sketch of tissue evidence from Exhibit A

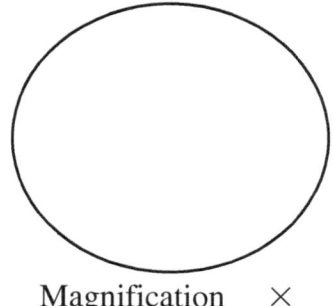

Magnification __×

Human Anatomy and Physiology

Your Name: _____   Today's Date: _____

Class: _____   Class Time: (Day) _____ (Time) _____

3. Positive ID on tissue? (What is the tissue type after examination and research?)

4. What are some likely causes of death? (Brainstorm with your team.)

B. Crime Scene #2—Young female victim (Case no. 8609447)

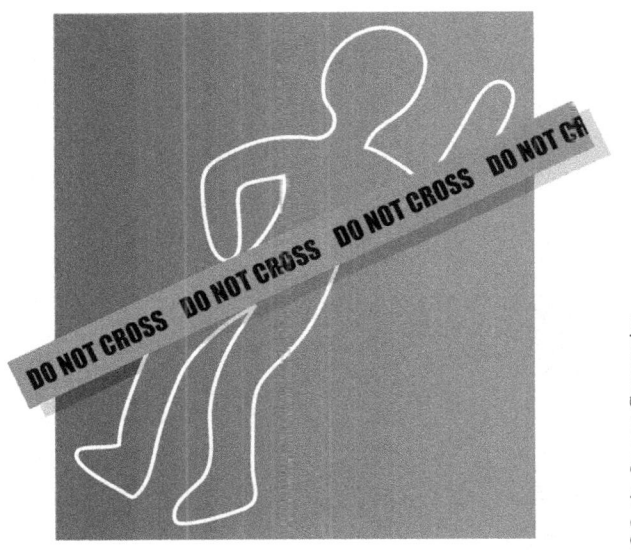

Description: A young female victim was discovered postmortem anterior side face up. The victim apparently had been reported as a missing person approximately 2 years ago and had not been seen since that time. A sterna medial stab wound was apparent approximately 4 inches in depth and 1.5 inches in width. The victim had no purse, identification, wallet, or otherwise. The weapon was recovered in a ditch near the victim's body. A tissue sample was taken from the knife's blade and is featured in Exhibit B (slide).

1. Probable origin of tissue? (Main muscle involved in fatal wound?)

Your Name: _____ Today's Date: _____

Class: _____ Class Time: (Day) _____ (Time) _____

2. Sketch of tissue evidence from Exhibit B

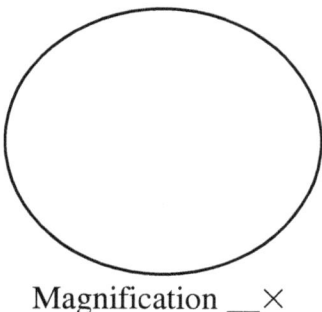

Magnification __×

3. Positive ID on tissue? (What is the tissue type after examination and research?)

4. Likely cause of death? (Brainstorm with your team.)

C. Crime Scene #3—Male victim

Description: Victim was found ventral side up approximately 5 hours postmortem. Victim had a 9mm handgun in left hand upon discovery. Upon closer examination, the handgun had not been fired. The victim had an apparent stab wound to the upper left leg. The weapon used to attack the victim was recovered as well as a tissue sample which is shown in Exhibit C (slide). No motive has been established at this time.

Your Name: _____ Today's Date: _____

Class: _____ Class Time: (Day) _____ (Time) _____

1. Probable origin of tissue? (Name of muscle(s) involved?)

2. Sketch of tissue evidence from Exhibit C

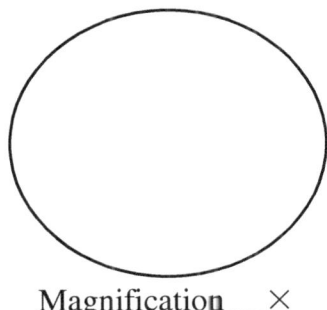

Magnification __×

3. Positive ID on tissue? (What is the tissue type after examination and research?)

4. Likely cause of death? (Brainstorm with your team.)

III. Follow-up questions and analysis

A. What are some similarities between the three crime scenes?

B. What are the three types of muscle tissue represented in the three crime scenes? What are the distinguishing physical characteristics of each muscle type that are notable during microscopic examination? (How can you tell them apart visually?)

Laboratory Exercise 10

Your Name: _____ Today's Date: _____

Class: _____ Class Time: (Day) _____ (Time) _____

C. How would you rate your knowledge of muscle anatomy and types of muscle tissue after completing this lab? (1 is least amount of knowledge and 10 is best)? Indicate your rating below.

_____
1   5   10

D. What did you learn from this lab?

Muscle Key

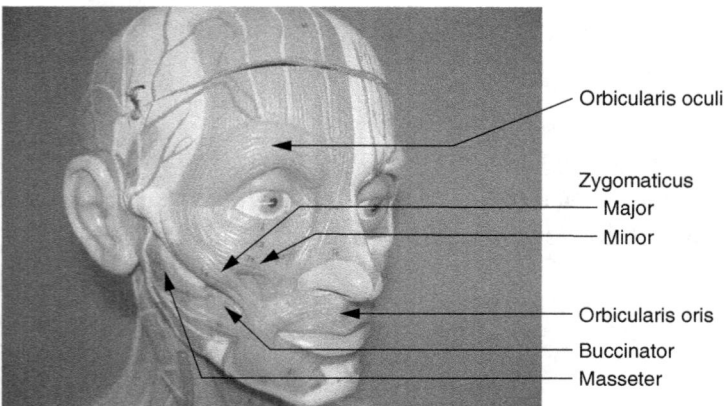

Your Name: _____ Today's Date: _____

Class: _____ Class Time: (Day) _____ (Time) _____

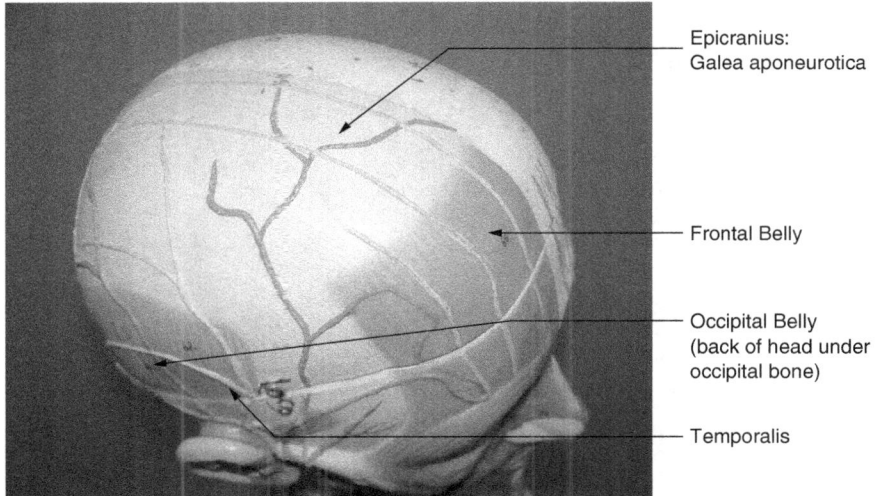

- Epicranius: Galea aponeurotica
- Frontal Belly
- Occipital Belly (back of head under occipital bone)
- Temporalis

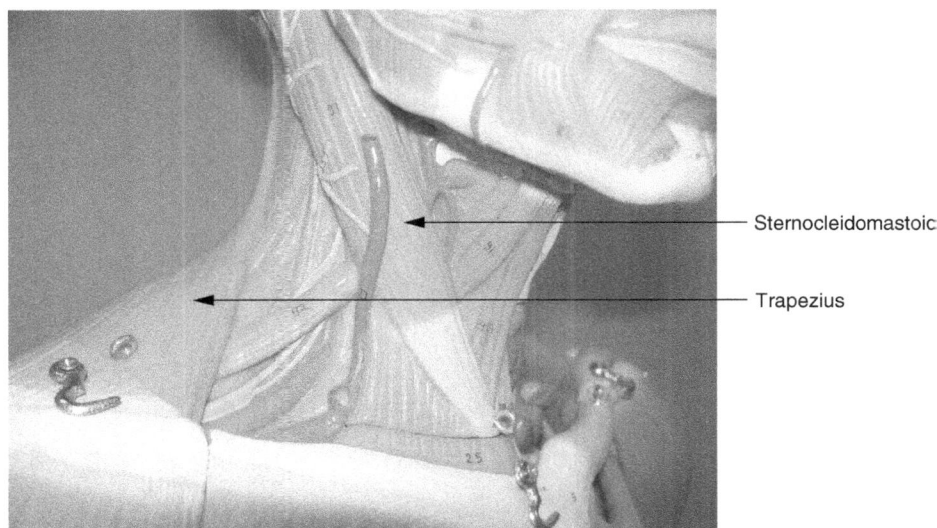

- Sternocleidomastoid
- Trapezius

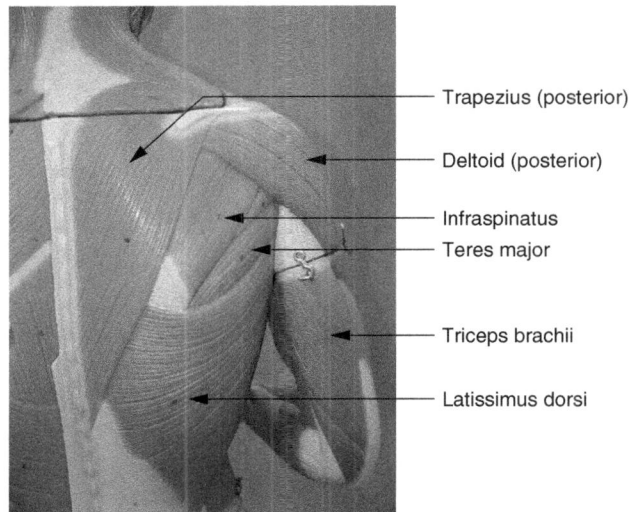

- Trapezius (posterior)
- Deltoid (posterior)
- Infraspinatus
- Teres major
- Triceps brachii
- Latissimus dorsi

Laboratory Exercise 10   91

Your Name: _____   Today's Date: _____

Class: _____   Class Time: (Day) _____ (Time) _____

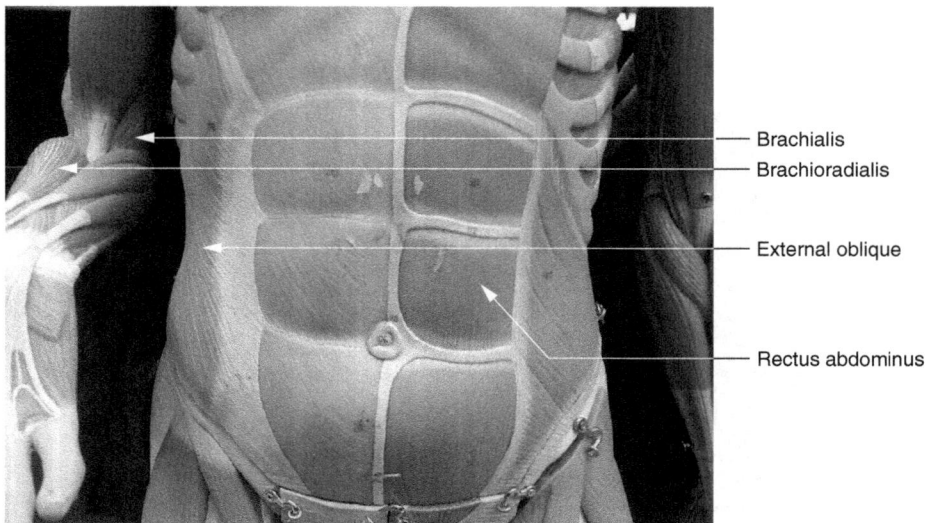

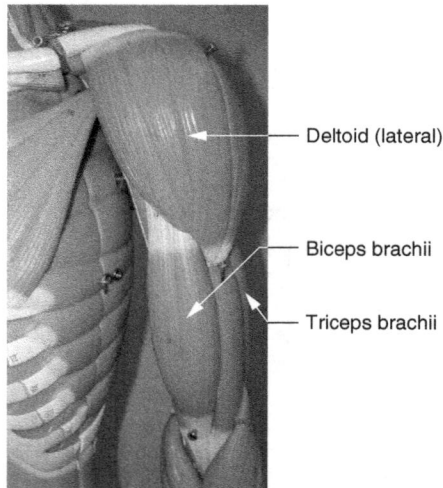

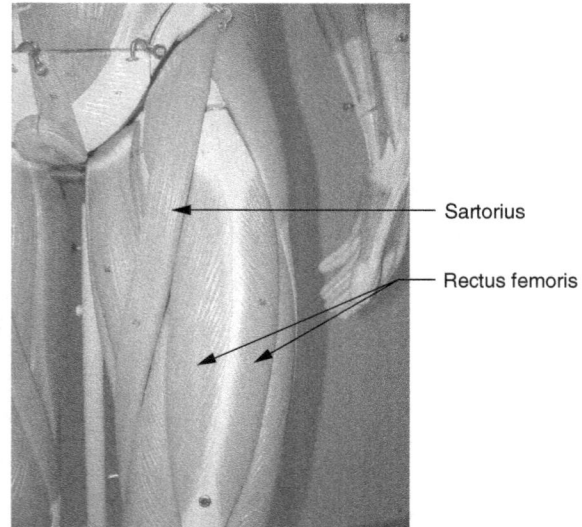

Human Anatomy and Physiology

Your Name: _____   Today's Date: _____

Class: _____   Class Time: (Day) _____ (Time) _____

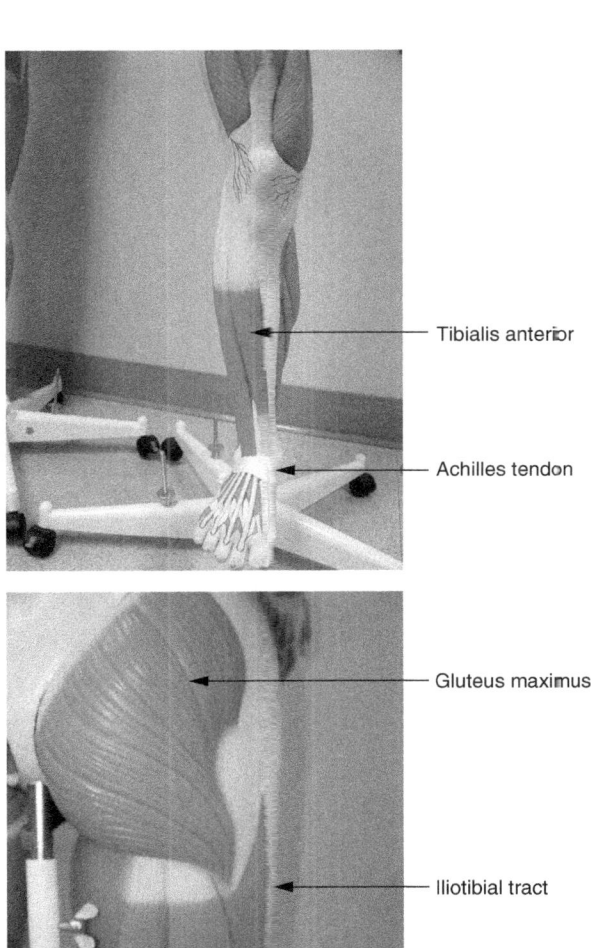

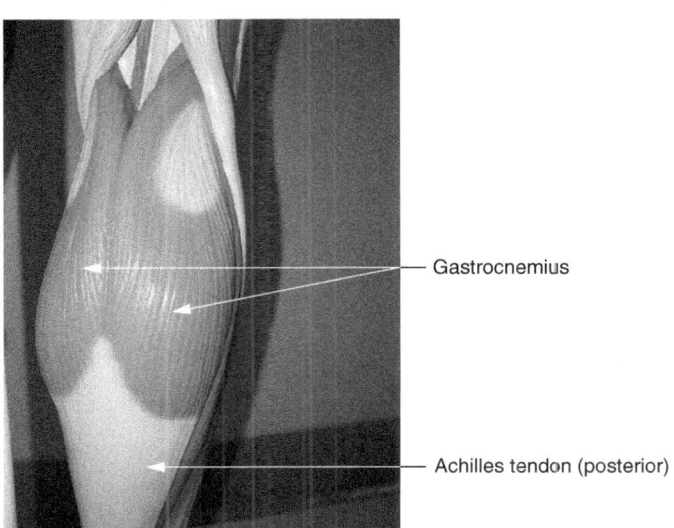

Laboratory Exercise 10

LABORATORY EXERCISE 11
The Nervous System Versus the Case of the Distracted Driver

Learning Objective(s)

The student will be able to:

- Identify a spinal cord neuron and name as many structures as visibly possible (axon, cell body, dendrites, etc.).
- Identify the major parts of the brain and explain what each area controls.
- Analyze a fatal driving accident and reenact various scenarios that physiologically explain how distractions affect reaction times while driving.

MATERIALS: neuron slide, microscope, brain model, ruler, stop-watch or second hand on watch

Your Name: _____   Today's Date: _____

Class: _____   Class Time: (Day) _____ (Time) _____

I. Introduction: The Nervous System

Working in small teams, complete the following together.

A. Introduction to Nerve Cells

How well do you know the human nervous system? How about the parts of your brain and what they do? In this laboratory exercise, you will have the opportunity to explore and better understand how things work.

Before we get rolling, do you know what the most basic unit of the nervous system is? The neuron is the cell unit of the nervous system. It conducts impulses both chemically through the use of (1) _____ and (2) electrically.

B. Exploration

Obtain a slide of spinal cord neurons. Draw the neurons as you see them. Be sure to label parts that you can actually see/recognize (cell body, nucleus, Nissl bodies, axon, Schwann cells, nodes of Ranvier, dendrites).

Neuron(s), _____ × power

II. Major Parts of the Brain

A. Find these structures on an anatomical model. Check off your progress as you go.

NOTE: The numbers on this chart should correspond with the numbers on the model.

Cerebral Hemispheres/ Endbrain
__ 1 Frontal lobe
__ 2 Parietal lobe
__ 3 Occipital lobe
__ 4 Temporal lobe
__ 5 Central sulcus
__10 corpus collosum

Diencephalon
__19 Thalamus
__21 Hypothalamus
__25 Pituitary gland

Cerebellum
__35 Cerebellum

Brain Stem
__36 Pons
__37 Medulla Oblongata

Laboratory Exercise 11

Your Name: _____   Today's Date: _____

Class: _____   Class Time: (Day) _____ (Time) _____

B. Labeling: Label all of the structures on the previous chart on one of the following diagrams.

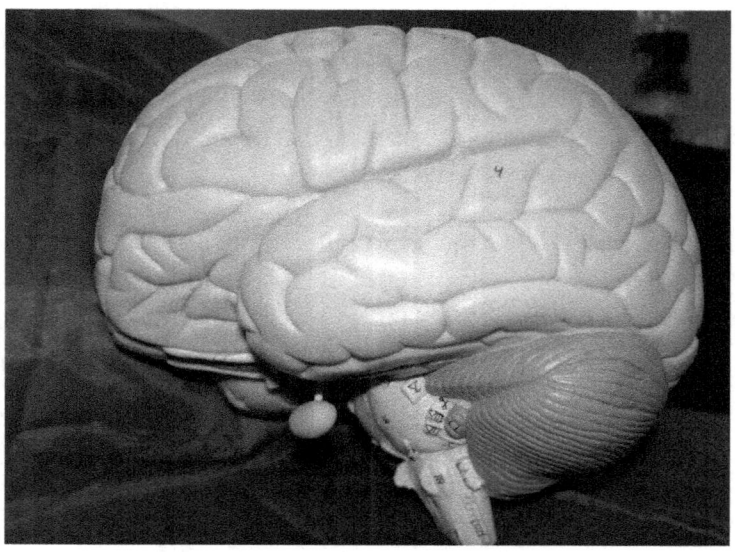

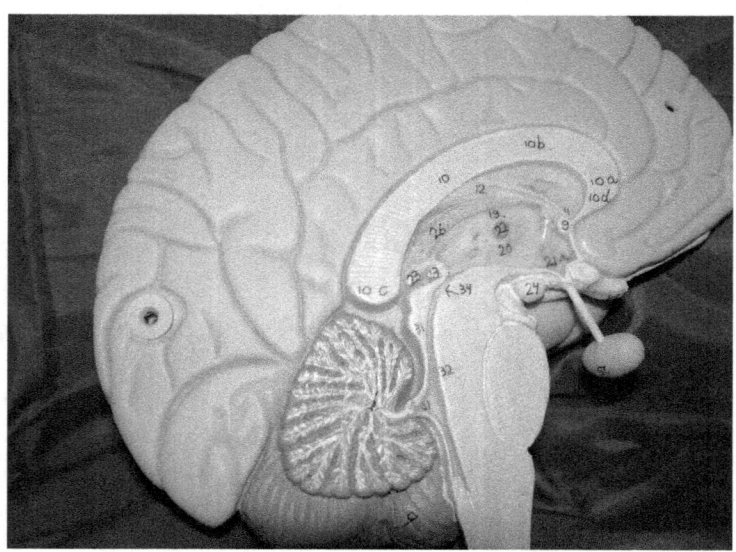

Human Anatomy and Physiology

Your Name: _____   Today's Date: _____

Class: _____   Class Time: (Day) _____ (Time) _____

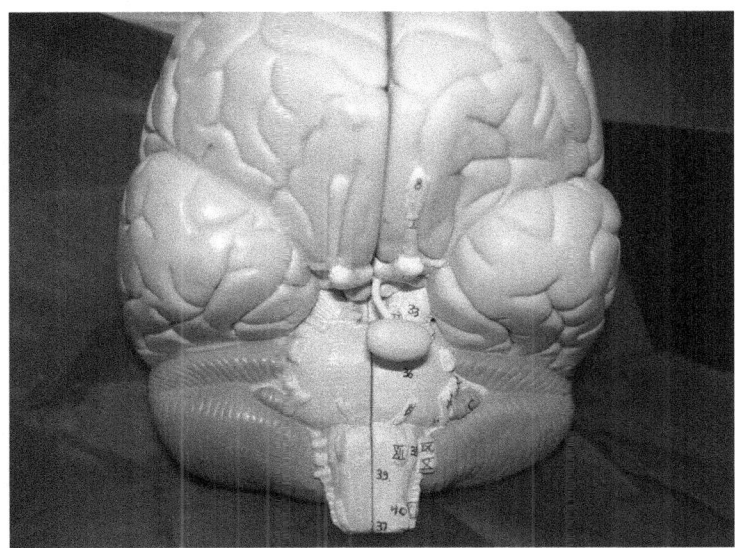

C. Functions: Use your text or other resources to explain the main functions of the major brain areas listed below. **NOTE:** This section may take some research time on the computer—you may want to complete it outside of laboratory time.

1. Frontal lobe

2. Parietal lobe

3. Occipital lobe

4. Corpus callosum

5. Thalamus

6. Hypothalamus

7. Pituitary gland

8. Pons

9. Medulla oblongata

10. Cerebellum

Laboratory Exercise 11

Your Name: _____ Today's Date: _____

Class: _____ Class Time: (Day) _____ (Time) _____

III. Analysis

Predict the effects of injuries to the following areas (you may want to use your text/notes for this):

1. Frontal lobes

2. Brainstem

3. Cerebellum

4. Occipital lobes

IV. Introduction: Reflexes and Reaction Times

A. How fast are your reflexes? Let's say, for instance, that you accidentally tip over a glass of water sitting near you. Are your reflexes so quick that you generally catch the glass before it spills? Or are you like many who attempt to catch the glass but inevitably watch as water goes everywhere? Physiologically speaking, why does it take time for your body to react? Write a brief explanation of how your nervous system works in a situation like this. Include the parts of the brain and nervous system involved in reflexes. If you do not know, use your text or other resources like the Internet to assist you in your understanding.

_____

_____

_____

_____

Your Name: _____ Today's Date: _____

Class: _____ Class Time: (Day) _____ (Time) _____

B. For busy Americans, the occurrence of traffic accidents appears to be on the rise. Apparently today's drivers are doing too many other things while driving! Research shows that distractions slow reaction time.

What are some things that you have observed drivers doing while driving that they probably shouldn't be doing? Or what are some things you yourself may have done while driving that wasn't the safest practice? Write your answer below.

_____

_____

_____

C. The Case of the Distracted Driver (carefully study the case).

On March 11, 2009, Brenda Griner's car left the road, resulting in her fatality. She was driving alone on a farm road at the time of the accident. The time of death was approximately 8:05 am. The car left the road and entered a canal where it quickly sank to the bottom, completely covering the car in water. One witness who saw the accident from a distant oncoming lane said she saw the car veer off the road at a high rate of speed and fly into the canal. She said she immediately dialed 9-1-1 and waited by the canal embankment. She mentioned that she cannot swim and did not attempt to dive in after the vehicle because of this reason.

Police records indicate that there appeared to be no other car involvement or foul play. The weather was marginal—not the best of conditions but not the worst either. There had been a slight mist most of the morning hours with partly cloudy conditions. Tire tracks indicated a speed of approximately 60 miles per hour. The speed limit on the farm road was indicated to be 45 miles per hour. No skid marks were apparent until the edge of the canal embankment. The distance from the road to the canal embankment was approximately 300 feet. No guard rail was apparent to slow the entry of the vehicle into the water.

The car was retrieved from the canal with Brenda's body behind the wheel. Other items collected from the car included a cell phone that was apparently on at the time of the accident and various makeup items including lipstick and mascara, which were both uncapped.

Autopsy results indicated a severe impact injury to the head and upper torso. Water from the canal was found in the victim's lungs. No blood alcohol or other drugs were found in the victim's body. The victim was pronounced dead at the scene.

Your Name: _____   Today's Date: _____

Class: _____   Class Time: (Day) _____ (Time) _____

Police have ruled out both mechanical failure and suicide. They have logged this as an accidental death. This is largely due to the fact that skid marks were found although they were late in the car's trajectory path. It's as if she didn't realize what was happening until it was too late.

1. Do you agree with the police report showing accidental death? Why or why not? What do you think happened? _____

   _____

   _____

   _____

   _____

   _____

   _____

2. If the vehicle was traveling at 60 miles per hour and traveled approximately 300 feet before reaching the canal embankment, how long did it take the driver to respond and apply the brakes? Explain how you arrived at your answer. (Show or explain the math)_____

   _____

   _____

   _____

3. Research supports a 1–2 second response time for driver's to respond in similar situations. What may account for the additional response time in Brenda's case?_____

   _____

   _____

D. Extension and Further Research

1. You can test reaction times with a very simple test. This activity requires three people.

Person #1 will give a "go" or "start" command.

Human Anatomy and Physiology

Your Name: _____ Today's Date: _____

Class: _____ Class Time: (Day) _____ (Time) _____

Person #2 will be the "dropper." This person will hold a ruler approximately 12 to 16" vertically above another person's hand and drop the ruler when Person #1 gives the command. **NOTE:** Be sure the ruler is dropped with 0" at the bottom end of the ruler (and not 12").

Person #3 is the catcher. As soon as possible, this person should catch the ruler between the thumb and forefinger. Look at the ruler to see where the ruler was caught. **NOTE:** 1" = 1 second reaction time, 2" = 2 seconds, etc.

Record reaction times below. Do three trials.

Trial	Seconds to React
1	
2	
3	

2. Now, repeat the same experiment but add a distraction of your choice to the activity. For example, catch the ruler while chewing gum or text messaging.

Distraction #1 _____

Trial with Distraction #1	Seconds to React
1	
2	
3	

a. Did the distraction affect reaction time? Explain what happened.

_____

_____

_____

b. How do you think adding another distraction will affect reaction times? (i.e., talking to someone and eating something with the opposite hand while catching the ruler) _____

_____

Laboratory Exercise 11

Your Name: _____ Today's Date: _____

Class: _____ Class Time: (Day) _____ (Time) _____

   3. Now, repeat the same experiment using two distractions.

   Distraction #1 _____ and Distraction #2 _____

   _____.

Trial with Distraction #1 and #2 simultaneously	Seconds to React
1	
2	
3	

   a. How did two distractions affect reaction times? Explain. _____

   _____

   _____

   b. What if someone was driving and had three or more distractions occurring simultaneously? What would you predict would happen to reaction times?

   _____

   _____

   _____

E. Conclusion

What have you learned from this lab? (i.e., What impressed you the most? What did you learn that was new? Anything that applies to your real life?) Compose your answer below.

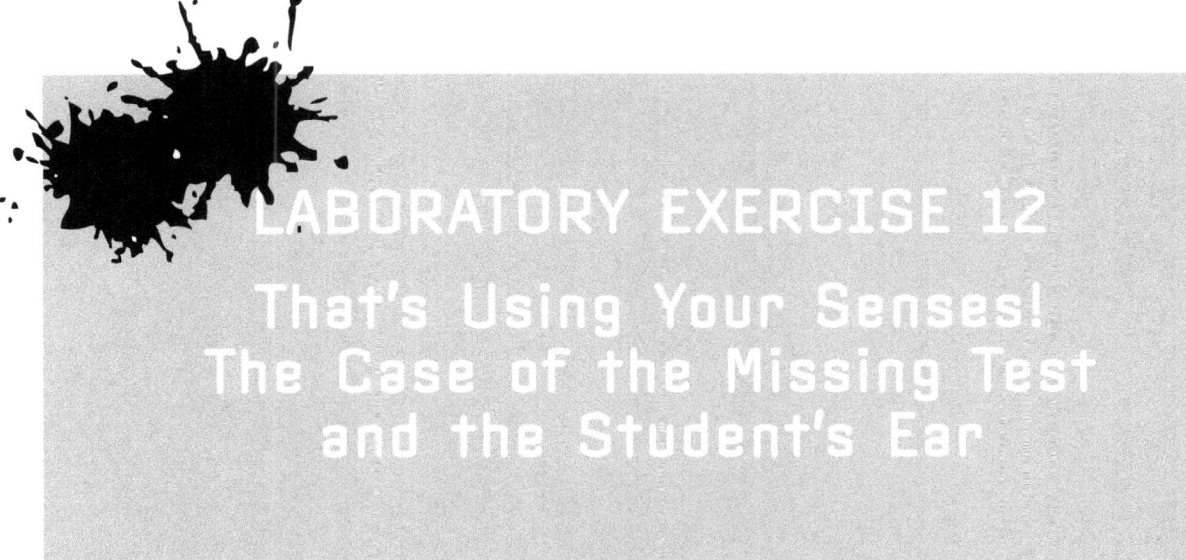

LABORATORY EXERCISE 12
That's Using Your Senses! The Case of the Missing Test and the Student's Ear

Learning Objective(s)

The student will be able to:

♦ Identify the major structures of the eye and ear on a human anatomical model.
♦ Use problem-solving skills to help solve a case involving an instructor's missing test and ear prints.

MATERIALS: eye model, ear model

Your Name: _____   Today's Date: _____

Class: _____   Class Time: (Day) _____ (Time) _____

I. Eye

A. **Introduction.** Sense organs like the eye and ear convey information from the outside world back to the brain. These organs are actually extensions of the nervous system. They are directly wired into the brain and allow us to automatically respond or conscientiously perceive our internal and external environments. A stimulus excites a sense organ, which then transduces the stimulus to an electrical (nerve) impulse. Sensory nerves transmit the impulse (sensation) to the brain to be perceived and acted upon. Ultimately, it is the brain that actually feels, sees, hears, tastes, and smells!

It is amazing how we see. The eyes refract (bend) and focus incoming light waves to special photoreceptor cells at the back of the eye. These photoreceptors cells are known as rods and cones. Nerve impulses from the stimulated photoreceptors are conveyed along visual pathways to the occipital lobes of the cerebrum, where vision sensations are perceived.

1. What are some things you already know about the eyes and vision? Brainstorm your ideas below.

   _____

   _____

   _____

2. In your own words, explain how you are able to see.

   _____

   _____

   _____

3. Where are photoreceptors cells located?

   _____

   _____

Laboratory Exercise 12

Your Name: _____ Today's Date: _____

Class: _____ Class Time: (Day) _____ (Time) _____

B. Eye Anatomy

1. Using an eye model and key, locate the following eye structures on the model. Check off your progress as you go along.

   __ superior rectus muscle   __ lens
   __ inferior rectus muscle   __ choroid
   __ lateral rectus muscle   __ retina
   __ superior oblique muscle   __ optic disc (blind spot)
   __ sclera   __ optic nerve
   __ pupil   __ iris
   __ iris   __ cornea

2. Key. Study this key and use it to help you identify the structures on the eye model.

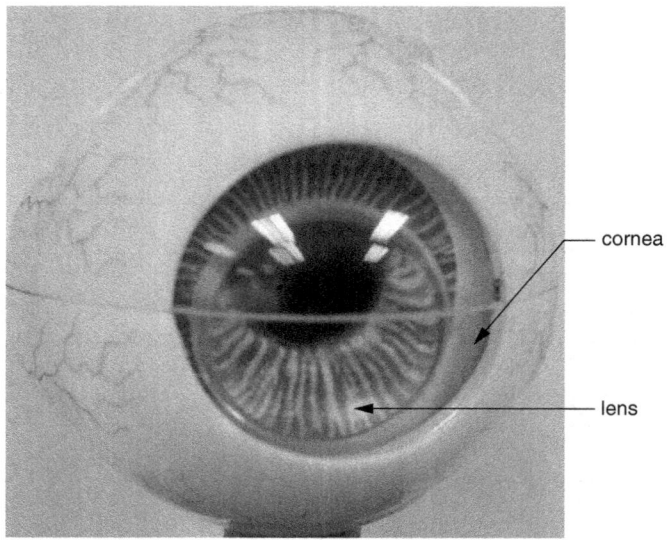

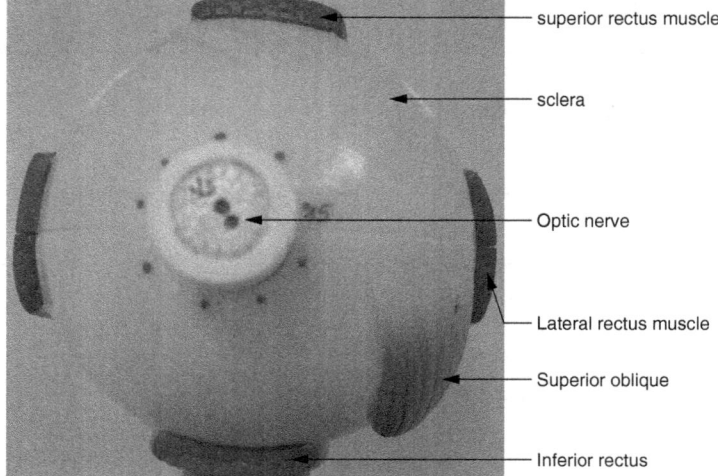

Human Anatomy and Physiology

Your Name: _____ Today's Date: _____

Class: _____ Class Time: (Day) _____ (Time) _____

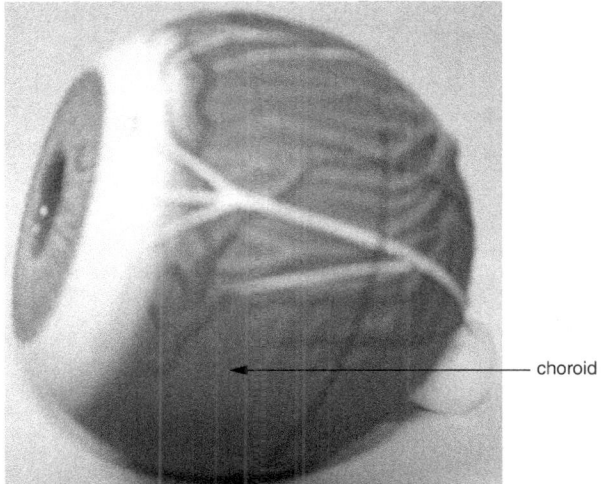
choroid

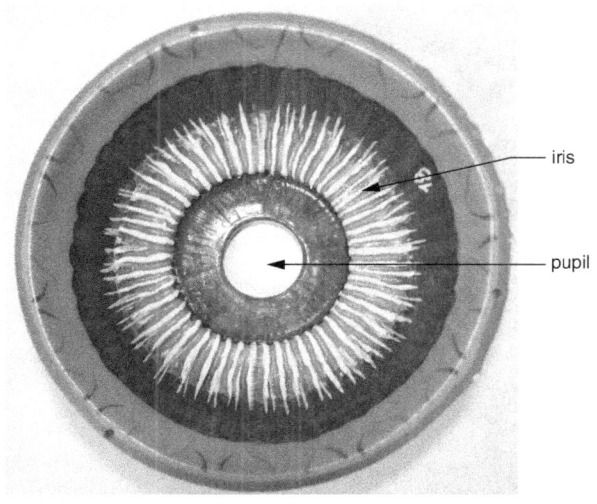
iris
pupil

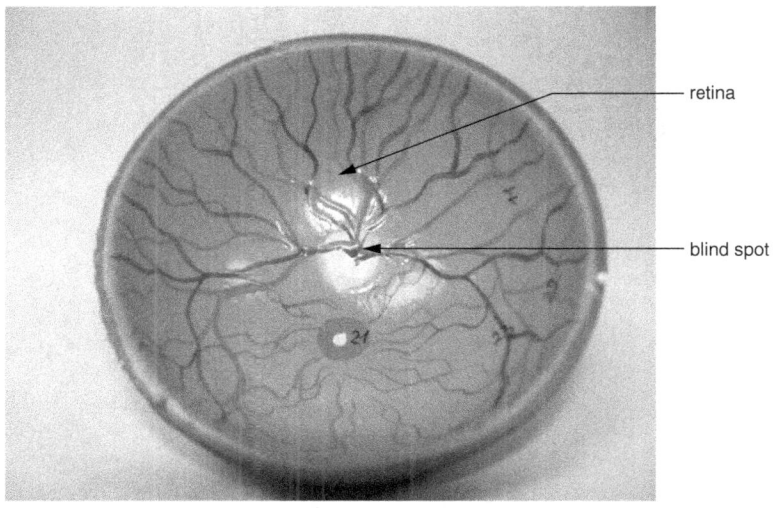
retina
blind spot

Laboratory Exercise 12

Your Name: _____   Today's Date: _____

Class: _____   Class Time: (Day) _____ (Time) _____

3. Labeling Practice. Challenge yourself to label these structures from memory.

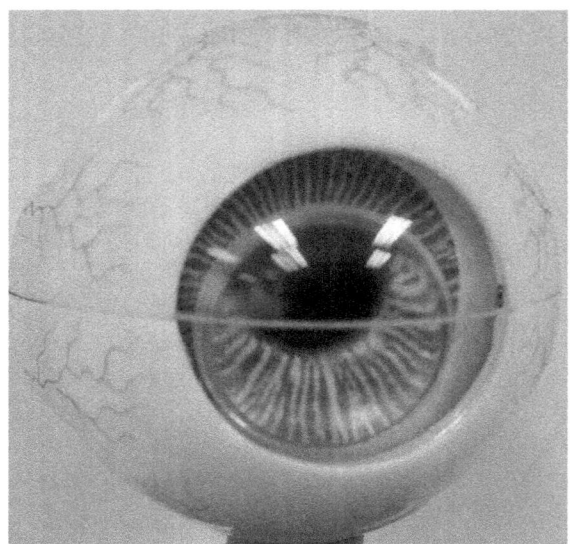

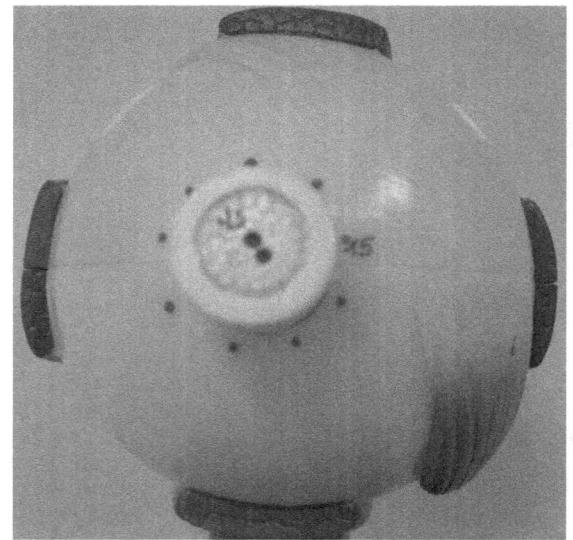

Human Anatomy and Physiology

Your Name: _____ Today's Date: _____

Class: _____ Class Time: (Day) _____ (Time) _____

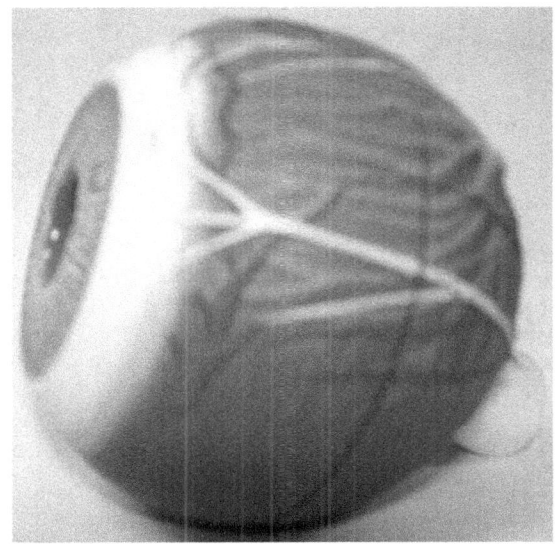

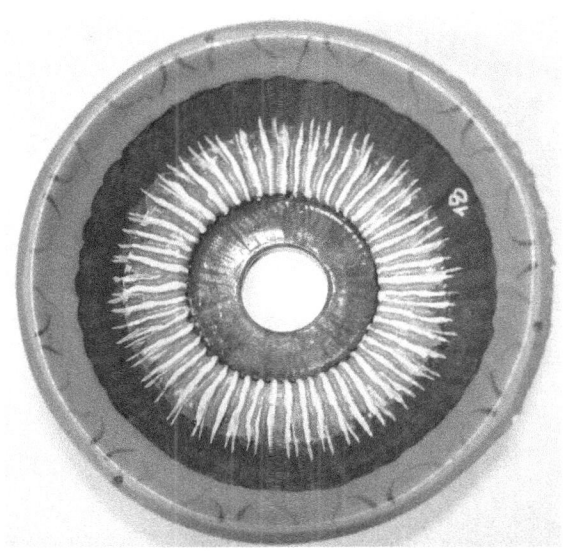

Laboratory Exercise 12

Your Name: _____ Today's Date: _____

Class: _____ Class Time: (Day) _____ (Time) _____

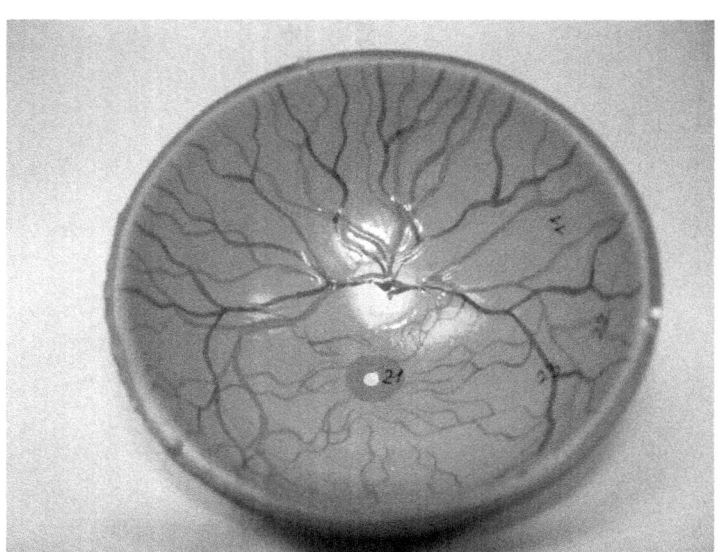

4. Explain the physiology of these structures:

   a. cornea

   b. iris

   c. pupil

   d. sclera

   e. choroid

   f. lens

   g. retina

   h. blind spot

   i. optic nerve

II. Ear

   A. Background on Ears and Crime Scenes

   Are any two fingerprints alike? Most forensics experts would answer a resounding "no." How about ear prints? Sound crazy? Think about it. The outside of the ear or (1) _____ is observably different on each person. There are different shapes, sizes, swirl patterns, lobe sizes, angles, and other features.

Your Name: _____ Today's Date: _____

Class: _____ Class Time: (Day) _____ (Time) _____

And it is not uncommon for a perpetrator to put an ear to a door or window prior to entering a crime scene in an effort to determine whether the area is occupied, or for an ear print to be left against a wall or other hard surface during a struggle or when a body is being positioned or moved. This evidence can be collected at the crime scene, using methods similar to those used for the lifting of fingerprints. A benefit to the collection of ear prints along with other crime scene evidence is in their use as confirmatory data. The legal system typically requires two different types of corroborative evidence in order to confirm placement of a suspect at a crime scene. While it is possible to "plant" fingerprints or even DNA material, it is difficult to intentionally place an ear print, particularly before ear prints become a common form of forensic identification, at a crime scene.

At present, there is a paucity of scientific evidence supporting the use of ear prints in forensic investigations. There has not been enough research evidence that ear prints are unique to each individual; there is a lack of systematization in the collection and analysis of ear print data; and there has not been widespread development or usage of automated ear print matching technology. Nonetheless, it is admissible evidence and along with other types of evidence may help lead to a positive ID.

Sample ear print from a generalized crime scene:

B. The Case of the Missing Test and the Student's Ear

It was time for end-of-the-year finals and the students in Mrs. Sock It-to-'Em's lab were all anxious. While most students were busy studying and preparing, word got around that some students were plotting shortcut strategies. Apparently, some were even entertaining the idea of trying to find out what the actual test questions would be ahead of time. How diabolical!

For Mrs. Sock-It-to-'Em, this "gossip" seemed too impossible to believe. Most all of her students were lovely, positive creatures who didn't seem capable of such mischief. There were, however, a few students who stood out in her memory for excessive verbal complaining. Unlike the hardworking types, they were always goofing off, running behind, and then complaining about how difficult everything always was.

Your Name: _____ Today's Date: _____

Class: _____ Class Time: (Day) _____ (Time) _____

So, Mrs. Sock-It-to-'Em remained on high alert status during the development of her final lab practicum. She decided to develop her questions the night before the exam. She diligently word processed them on the computer and printed them out. Then, in order to prevent even the possibility of computer theft, she carefully deleted all traces of the file from her computer. She then placed the printout in a dedicated, but unmarked light blue folder on the corner of her desk. Satisfied, she locked up her office and left for the night.

The next morning, she came in bright and early. She passed several of her students, Distracted Dan and Nervous Nelson, on the way to her office. Both said hello as they quickly passed by. Distracted Dan was chewing gum or something—his hello was delayed. As Mrs. Sock-It-to-'Em rounded the corner to her office, she saw Effortless Evelyn, who had requested an appointment and was waiting for her. She cheerfully greeted her and asked Effortless Evelyn to come on in. She unlocked her office door and entered. Mrs. Sock-It-to-'Em noticed the blue folder still on the corner of her desk where she had left it. She invited Effortless Evelyn to have a seat while she went to the breakroom to pour a quick cup of coffee. She promptly re-entered the office to meet with Effortless Evelyn regarding a few questions on one of the labs done earlier in the semester. It was a brief meeting lasting only several minutes. Effortless Evelyn had her questions answered and left.

It was now time to copy the tests for this morning's final. Mrs. Sock-It-to-'Em was getting ready to head to the copy machine when there was a sudden knock at the door. Another student, Frantic Fran, needed some last minute help with some questions. Mrs. Sock-It-to-'Em asked her to come in and have a seat. Just as she did, there was another knock at the door. An unfamiliar student or for lack of better wording, a Strange Student, asked for directions to another instructor's office. Dropping what she was doing, Mrs. Sock-It-to-'Em was happy to help. She asked Frantic Fran to wait until she returned. She then quickly escorted the student down the hall to the faculty member's office. She then decided to hit the restroom before heading back to her office. She came back to meet with Frantic Fran. Frantic Fran was very talkative and was describing how difficult it was to study because she had ten kids all below the age of 12 and worked four part-time jobs while going to school. Being discreet, Mrs. Sock-It-to-'Em decided to shut her office door (so no one else could over hear the conversation) as she continued listening to Frantic Fran's dilemma.

While they continued talking, they both heard a strange noise outside the door. It sounded as if somebody had brushed up against the door. Was someone eavesdropping or was she just overreacting? After all, there was a "buzz" about students out to get test questions ahead of time. Mrs. Sock-It-to-'Em decided to look outside the door. When she opened the door, two students were scurrying around the corner. She didn't recognize them, although she thought one had a black hooded sweatshirt.

Your Name: _____ Today's Date: _____

Class: _____ Class Time: (Day) _____ (Time) _____

Despite the distraction, she continued her meeting with Frantic Fran but asked her to re-schedule because she needed to finish making test copies (test time was fast approaching). Frantic Fran agreed and left.

Before she could continue, there was yet another knock at the door—another lost student. She left the office momentarily to show the student the way. On the way back to her office, she passed Distracted Dan and Nervous Nelson again. She wondered to herself why they were back in this end of the building again but dismissed the thought.

Next, she ran into another student of hers—Irresponsible Iris—who wanted to know when the final exam would be, but in reality no matter the date or time, she couldn't be there. When Mrs. Sock-It-to-'Em finally re-entered her office a few minutes later to continue her copying efforts, the blue folder was missing. Nothing else appeared to be missing or disturbed. She did notice a partial muddy shoeprint outside her office **(see Exhibit A)**. There also appeared to be an ear smudge mark (ear print) on the outside of the door's glass window. Upon close examination, the top of the ear print was "pointy" not rounded. The ear print was fairly "tall" on the door, approximately at the level of at least 5′10″ plus. Additionally, there was a piece of candy on the ground. It was red and round **(see Exhibit B)**.

Despite her best efforts, the test questions had come up missing the morning of the exam! Had she simply misplaced them or had they actually been stolen? She searched her desk, files, and entire office. She turned the place upside down—no test questions surfaced. All the work that she had done in advance was wasted. She would have to start over—yet there no time to start over. This did not set well with her. She has decided to ask for assistance in getting to the bottom of this matter. You have been selected to help. Use your sleuth skills to help Mrs. Sock-It-to-'Em get to the bottom of this situation and find the missing test.

1. Here is a list of suspect students:

Effortless Evelyn was noted by other faculty and staff members as "hanging around the hallway" apparently waiting for an appointment with her teacher earlier this morning; generally an "A" student known by students as making it all look easy to accomplish; about 5′ 6″ tall with long, light brown hair and blue eyes; dressed in a purple shirt and jeans with new-looking white and purple tennis shoes

Frantic Fran, small frame and about 5′ 2″ with brown hair; described by others as always in a hurry rushing from one destination to the next; unorganized but trying to make it through school; sometimes wears inappropriate clothes to class like skimpy halter tops; usually wears very old, dirty tennis shoes (typically Keds brand).

Your Name: _____   Today's Date: _____

Class: _____   Class Time: (Day) _____ (Time) _____

Strange Student, short statue and muscular with brown hair; never seen before; apparently investigating the possibility of going to college and in need of advising.

Distracted Dan, about 5′ 8″ with fair skin and blonde hair; regarded by other students as never focused on school and always distracted by something; not particularly well liked by his classmates; always chewing gum or eating candy in class; famous for wearing a sweatshirt with a hood even when it was warm outside.

Nervous Nelson, tall, thin stature approximately 5′ 11″ with very short brown hair; always wears a plain white t-shirt; regarded by his classmates as nervous because he is always chewing his fingernails, biting the skin around his nails, or fidgeting nervously; has been known to annoy others who sit near him because he nervously wiggles his leg non-stop for no apparent reason, which wiggles the desks of others nearby; last class period (1 day previous) he was wiggling his leg and candy dropped out of his jean pocket and went everywhere on the classroom floor (he left it there). The candy was extracted from the classroom **(see Exhibit C)**.

Lost student, small-framed black female; no memorable characteristics other than short hair and big red heart-shaped earrings; no description or photos available.

Irresponsible Iris, small stature standing approximately 5′ 4″ with dark blonde hair; students describe her as always having an excuse and never taking responsibility for her own mistakes/ actions.

2. Each suspect has also been asked to submit an ear photo:

Effortless Evelyn

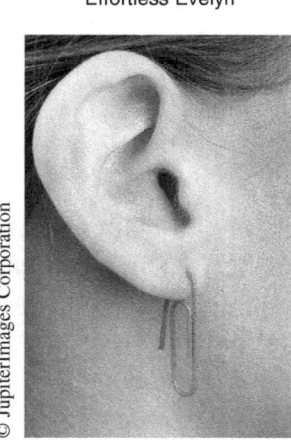

Frantic Fran

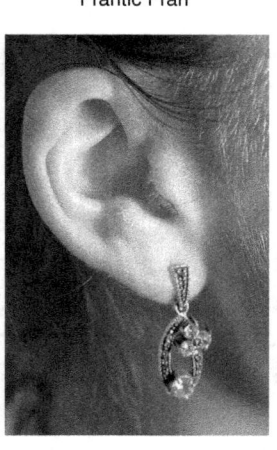

Human Anatomy and Physiology

Your Name: _____ Today's Date: _____

Class: _____ Class Time: (Day) _____ (Time) _____

Strange Student Distracted Dan

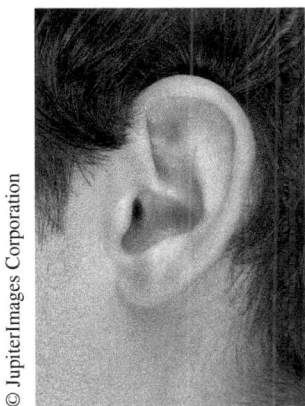
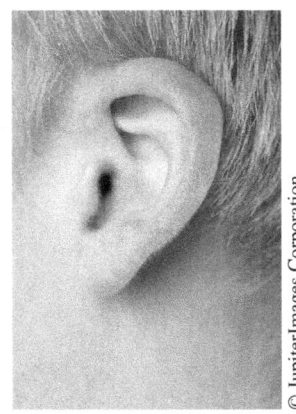

Nervous Nelson Irresponsible Iris

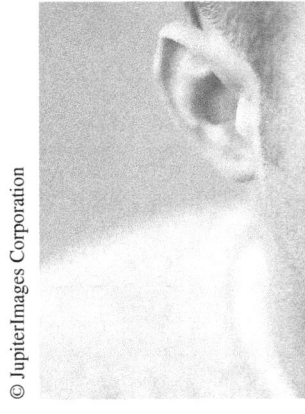
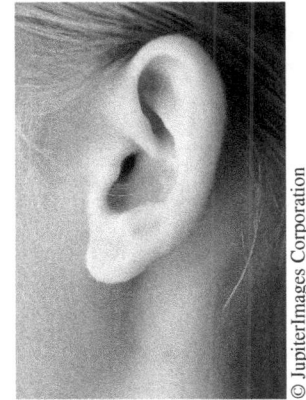

3. Exhibits:

Exhibit A   Exhibit B

Laboratory Exercise 12

Your Name: _____   Today's Date: _____

Class: _____   Class Time: (Day) _____ (Time) _____

Exhibit C

1. Analysis

    a. How are the **suspects'** outer ears physically different from one another's?

    _____

    _____

    b. How are your **group members'** outer ears different from one another's?

    _____

    _____

    _____

2. Follow-up Questions:

    a. What is your initial hypothesis about the disappearance of the test? (i.e., foul play involved or not? If yes, what possibly happened? Who may have taken the test?)

    _____

    _____

    _____

    _____

Human Anatomy and Physiology

Your Name: _____ Today's Date: _____

Class: _____ Class Time: (Day) _____ (Time) _____

_____

_____

_____

    b. What key fact or evidence lead supports your group's hypothesis?

_____

_____

_____

_____

    c. Can the ear print evidence alone prove who did it? Why or why not?

_____

_____

_____

C. Ear Anatomy

  1. Find the following structures on an anatomical model. Check off your progress as you go along:

- __ pinna
- __ external auditory canal
- __ tympanic membrane (eardrum)
- __ ossicles:
  - __ malleus
  - __ incus
  - __ stapes
- __ middle ear cavity
- __ Eustachian tube
- __ auditory nerve
- __ cochlea
- __ semi-circular canals

Your Name: _____ Today's Date: _____

Class: _____ Class Time: (Day) _____ (Time) _____

2. Ear Key. Use this key to help your identify key structures.

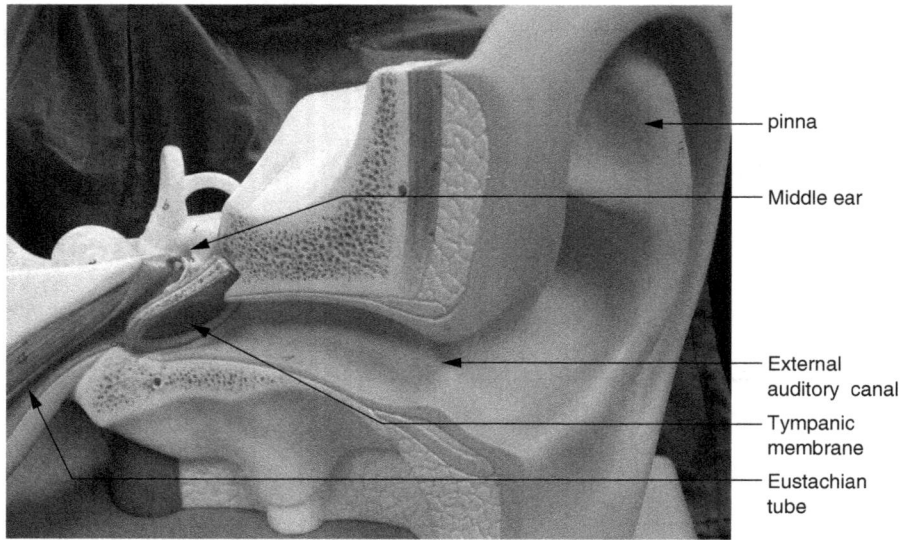

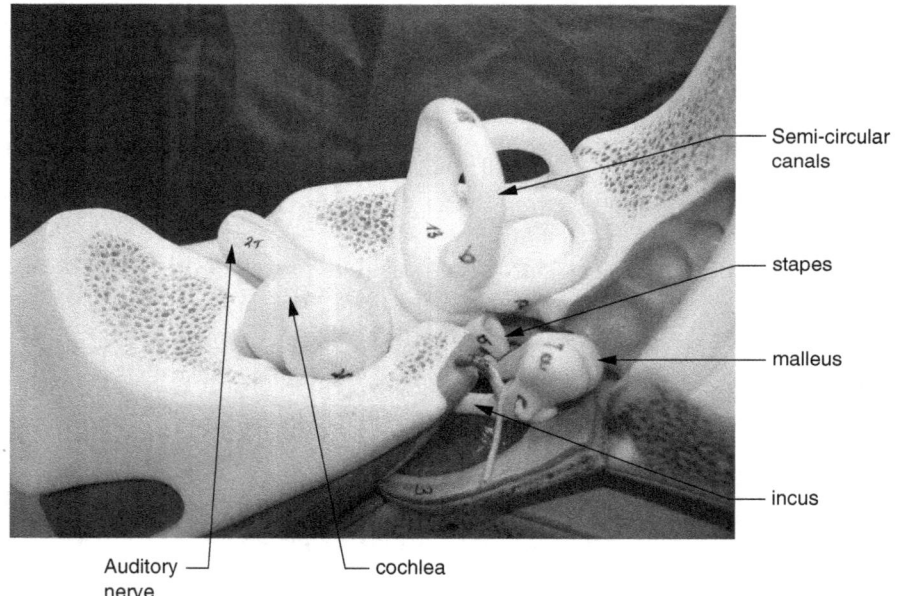

Your Name: _____ Today's Date: _____

Class: _____ Class Time: (Day) _____ (Time) _____

3. Label the ear. Challenge yourself not to refer to the key; try from memory.

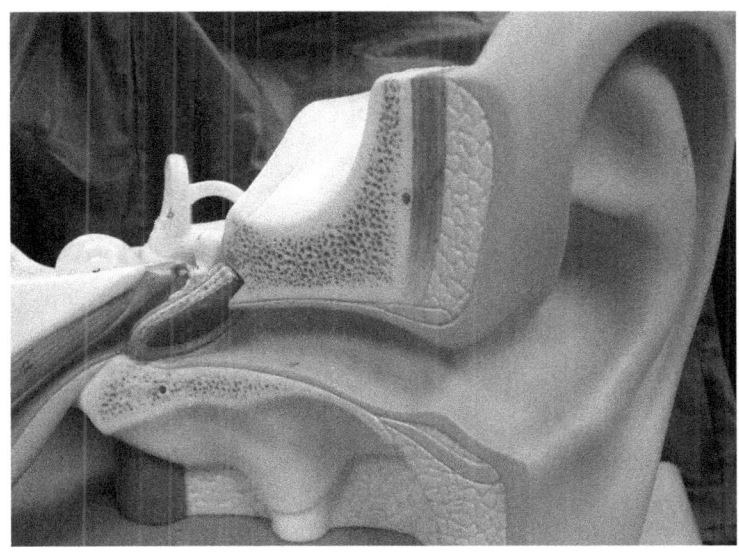

4. Explain the physiology of these structures:

   a. pinna

   b. external auditory canal

   c. tympanic membrane

   d. ossicles

Laboratory Exercise 12

Your Name: _____ Today's Date: _____

Class: _____ Class Time: (Day) _____ (Time) _____

      e. middle ear

      f. Eustachian tube

      g. cochlea

      h. semi-circular canals

      i. auditory nerve

D. Summary and Conclusion

In your own words, describe what you learned from today's lab. Please include what you found to be most interesting as well.

CPSIA information can be obtained
at www.ICGtesting.com
Printed in the USA
LVHW052017150819
627779LV00002B/3/P